Cambridge Atmospheric and Space Science Series

EDITORS

TITLES IN PRINT IN THE SERIES

M. H. Rees
Physics and chemistry of the upper atmosphere

R. Daley
Atmosphere data analysis

J. K. Hargreaves
The solar–terrestrial environment

J. R. Garratt
The atmosphere boundary layer

S. Sazhin
Whistler-mode waves in a hot plasma

S. P. Gary
Theory of space plasma microinstabilities

I. N. James
Introduction to circulating atmospheres

T. I. Gombosi
Gaskinetic theory

M. Walt
Introduction to geomagnetically trapped radiation

B. A. Kagan
Ocean–atmosphere interaction and climate modelling

D. Hastings and H. Garrett
Spacecraft–environment interactions

J. C. King and J. Turner
Antarctic meteorology and climatology

T. E. Cravens
Physics of solar system plasmas

J. F. Lemaire and K. I. Gringauz
The Earth's plasmasphere

T. I. Gombosi
Physics of space environment

J. Green
Atmospheric dynamics

G. E. Thomas and K. Stamnes
Radiative transfer in the atmosphere and ocean

I. G. Enting
Inverse problems in atmospheric constituent transport

R. D. Hunsucker and J. K. Hargreaves
The high-latitude ionosphere and its effects on radio propagation

N. Meyer-Vernet
Basics of the solar wind

V. Y. Trakhtengerts and M. J. Rycroft
Whistler and Alfvén mode cyclotron masers in space

R. W. Schunk and A. F. Nagy
Ionospheres: Physics, plasma physics, and chemistry

M. C. Serreze and R. G. Barry
The Arctic Climate System
Second Edition

An Introduction to Space Plasma Complexity

Tom Tien Sun Chang
Massachusetts Institute of Technology

CAMBRIDGE
UNIVERSITY PRESS

CAMBRIDGE
UNIVERSITY PRESS

32 Avenue of the Americas, New York, NY 10013-2473, USA

Cambridge University Press is part of the University of Cambridge.

It furthers the University's mission by disseminating knowledge in the pursuit of education, learning, and research at the highest international levels of excellence.

www.cambridge.org
Information on this title: www.cambridge.org/9780521642620

First published 2015

Printed in the United States of America

A catalog record for this publication is available from the British Library.

Library of Congress Cataloging in Publication data
Chang, Tien Sun
An introduction to space plasma complexity / Tom Tien Sun Chang, Massachusetts Institute of Technology.
pages cm. – (Cambridge atmospheric and space science series)
Includes bibliographical references and index.
ISBN 978-0-521-64262-0 (hardback)
1. Space plasmas. 2. Astrophysics. I. Title.
QC809.P5C43 2014
523.01–dc23 2014045083

ISBN 978-0-521-64262-0 Hardback

To my wife, Monica, for her continued understanding, love, and encouragement.
To my children and their spouses, Adele, Richard and Michelle, and Alexandra and Matthew, for their love and cheerleading.
To my grandchildren, Oscar and Emerson, for their relaxing and joyful innocence.

Contents

Preface

Everywhere one peers into the space plasma environment – whether it is the ionosphere, the magnetosphere, the heliosphere, the interstellar medium, or beyond – there is complexity. This introductory text is written with the hope that after perusing its contents, students and researchers will be able to step readily into the arena of this fascinating and developing field – space plasma complexity. I wish to thank Michael Rycroft, who enticed me to initiate this venture when I was visiting the International Space Science Institute (ISSI) in Bern, Switzerland. My attention was distracted by the many new developments in the field in the intervening years and it is only during these past few months that I finally had the time to try to complete the project. Matt Lloyd at Cambridge University Press has been extraordinarily patient throughout the interim. I am grateful to him for allowing me to take such a long time to accomplish this most pleasant task. This brief monograph does not pretend to cover every topic or methodology in the theory of complexity nor does it attempt to provide an adequate description of the physics of space plasmas. My guiding principle throughout has been to produce an entry-level monograph such that anyone with a college degree in science or engineering may read, comprehend, and appreciate the richness of the subject matter. I hope that I have achieved this goal.

The book starts with an introduction of the phenomena of criticality for systems at thermodynamic equilibrium. The motivation here is to develop the basic ideas of complexity and introduce some of the special terminologies that have become popular in the current developments of the subject. The concepts of critical dynamics such as forced and/or self-organized criticality (FSOC) then follow with relevant space plasma applications. The discussions evolve into the description of the physics of complexity involving the dynamic interactions of coherent structures and the basic ideas of coarse-grained dissipation. The intermittent fluctuations resulting from such ideas and concepts are studied and quantified using the analyzing techniques of wavelet transforms, probability distribution functions of scale-dependent incremental changes, structure functions, partition functions of singular measures, generalized dimensions, and rank-ordered multifractal analysis (ROMA). A brief introduction of the dynamic renormalization group (DRG) then follows, with the aim to delineate the differences between criticality and multifractality and to provide a possible interconnection among such processes through the concepts of local and global invariants and the phenomenon of crossovers. A toy model exemplifying these abstractions concludes this monograph.

It is with great pleasure I express my gratitude for the continued support from my colleague, Cheng-chin Wu, who believed in my often oddball thoughts and provided me much needed collaboration throughout the past score of years since I began to develop my interest in this alien field of research. The book is in many ways the product of our cooperative efforts.

The content of this treatise is enriched by the generosities of the authors and publishers who kindly allowed me to quote and reprint the contents of their original research results, which constitute a significant portion of the selected examples in this monograph.

I am blessed to have an understanding family who always supported my single-minded activity of academic research. My heartfelt thanks to Monica, who has been the sustained thrust throughout my research career and who provided the essential reminders to complete this seemingly never-ending writing project. Adele, Richard, and Alexandra have been persistent cheerleaders.

It has been an exhilarating experience to complete this manuscript on crossover and multifractals – one that I shall always savor in my intermittent venture into the fluctuating criticalities.

MIT Kavli Institute
Cambridge, MA

Acknowledgments

I am grateful to my many colleagues and coresearchers, including Olga Alexandrova, Hannes Alfvén, Mats André, Vassilis Angelopoulos, Roger Arnoldy, Maha Ashour-Abdalla, Dan Baker, Supriya Banerjee, Bamandas Basu, Sunanda Basu, John Belcher, Roberto Bruno, Jörg Büchner, S. Chandrasekhar (Chandra), Jerry Chao, Sandra Chapman, Frank Cheng, Abe Chian, Giuseppe Consolini, Bruno Coppi, Geoff Crew, Chris Dum, Marius Echim, James Ernstmeyer, Carl-Gunne Fälthammar, Guru Ganguli, Carl Garland, Peter Gary, Tohru Hada, Alex Hankey, Brian Hassard, Mike Hesse, Bogdan Hnat, Mary Hudson, Bengt Hultqvist, Jack Jasperse, Jay Johnson, Charlie Kennel, John Keyes, Paul Kintner, Margy Kivelson, Alex Klimas, Dave Klumpar, Hannu Koskinen, Masha Koznetsova, Hervé Lamy, Lou Lee, Joe Lemaire, Manfred Leubner, Geof Ludford, Tony Lui, Rickard Lundin, Kris Lynch, Wieslaw Macek, Eckart Marsch, Hiroshi Matsumoto, Bill Matthaeus, Sasha Milovanov, Jeff Nicoll, Stan Olbert, Yoshi Omura, Gene Parker, Bill Peterson, Viviane Pierrard, John Podesta, John Retterer, Ilan Roth, Nat Rynn, Walt Sartory, Yaacov Shapira, Surja Sharma, Bob Schunk, Gene Stanley, David Sundkvist, Sunny Tam, Dave Tetreault, Rudy Treumann, Bruce Tsurutani, Vadim Uritsky, Dimitri Vassiliadis, Zoltan Vörös, Dimitri Vvedensky, Nick Watkins, Jim Weygand, Ken Wilson, Dave Winningham, S. T. Wu, Fareed Yasseen, Peter Yoon, Emiliya Yordanova, Jim Young, Gary Zank, and Lev Zeleny, who helped to mold my thinking and improve my understanding of the phenomena of criticality, multifractals, and space plasmas.

The National Science Foundation, the Air Force Office of Scientific Research, and the National Aeronautics and Space Administration of the U.S. government and most recently the STORM project sponsored by the European Community have provided the vital monetary support of the various activities related to my complexity research. In addition, ISSI's kind hospitality of hosting me for two serene summers in Bern for this project is gratefully acknowledged. The MIT Kavli Institute for Astrophysics and Space Research has provided me an ideal environment to conduct basic research covering the greater portion of my academic career. I am very appreciative of its continued support.

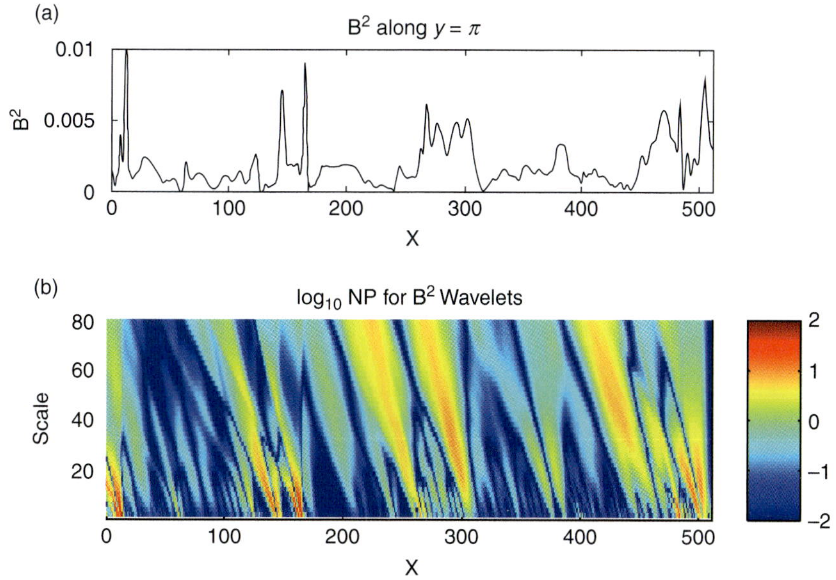

Figure 1.2. Local Intermittency Measure (*LIM*) or Normalized Power (*NP*) of the fluctuations of the magnetic energy density B^2 in arbitrary units at constant values of y and t obtained from a 2D (512×512) MHD homogeneous turbulence simulation using the Haar transform. The x-axis and scales are in units of grid spacing.

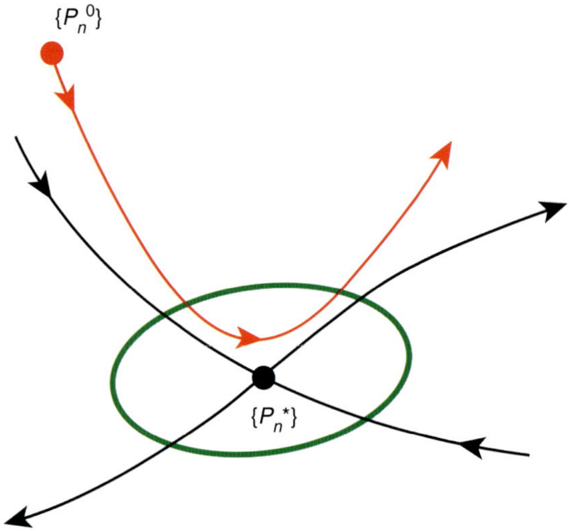

Figure 1.3. Parameter space of a dynamic renormalization group (RG). A typical (red) RG trajectory initiates from some point $\{P_n^0\}$, skirts close to and remains under the influence of the (green elliptical) domain of attraction of the partially repulsive (saddle) fixed point $\{P_n*\}$ before migrating away. Also shown are (black) trajectories (separatrices) that emerge or terminate at the fixed point.

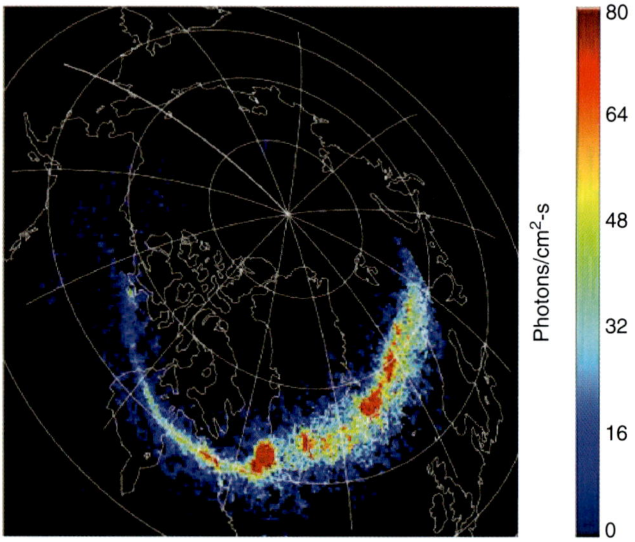

Figure 2.7. A sample of global auroral distribution from the UVI imager of the Polar spacecraft. (From Lui et al. 2000.)

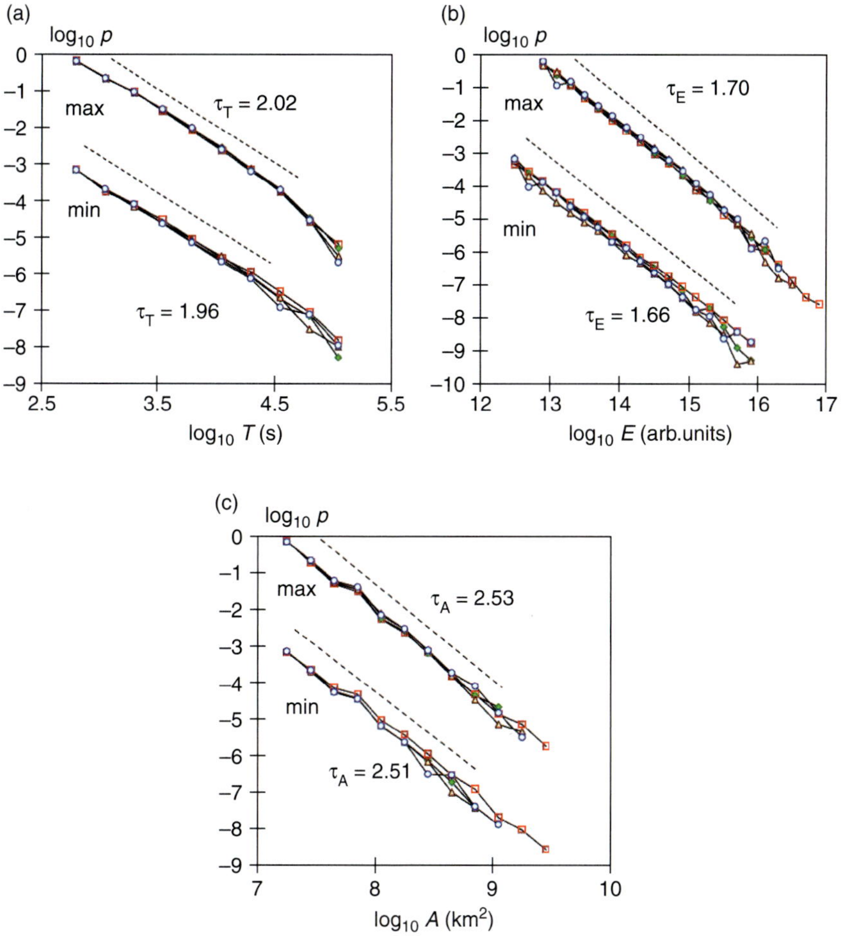

Figure 2.13. Probability distributions of spatiotemporal coronal avalanche lifetime T, EUV emission flux E, and peak area A, for solar max and min at different thresholds w_a (with different colors and geometrical shapes). The solar min distributions are shifted downward for comparison. (From Uritsky et al. 2007.)

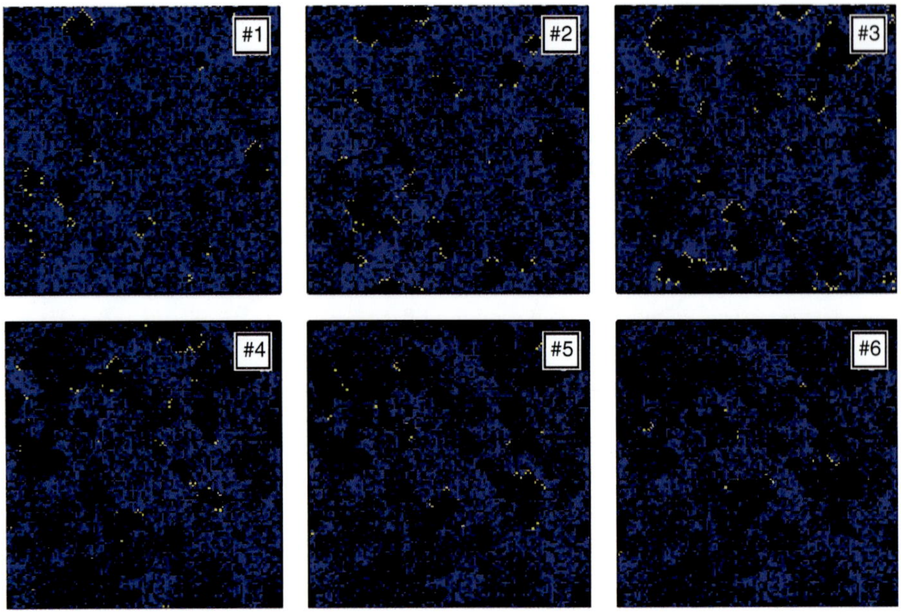

Figure 2.17. Sequence of snapshots of the RFFM cellular automaton. Time interval between snapshots is 10 timesteps. Stable sites (black background), unstable sites (violet diffused patches), and relaxing sites (yellow sharp spots). Note the localized and intermittent character of the relaxation processes. (From Consolini and De Michelis 2001.)

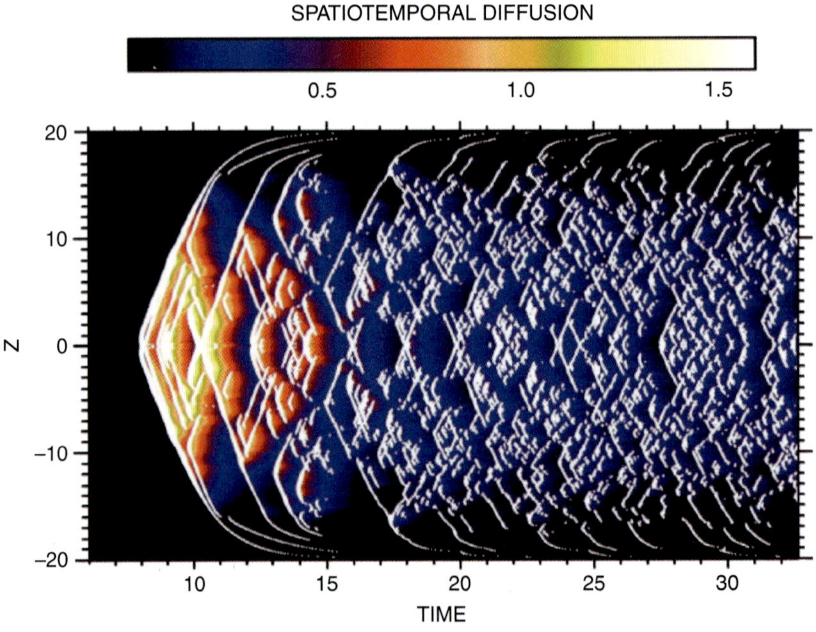

Figure 2.24. Spatiotemporal evolution of the diffusion coefficient $D(z,t)$ in arbitrary units for the Lu-Klimas field reversal model during the initial stage of development of an unloading process. The color scale of the figure indicates the intensity of the diffusion coefficient in arbitrary units. A white dot is placed at each point where the critical level for Q is reached and $Q(z,t) = D_{\max}$. (From Uritsky et al. 2001b.)

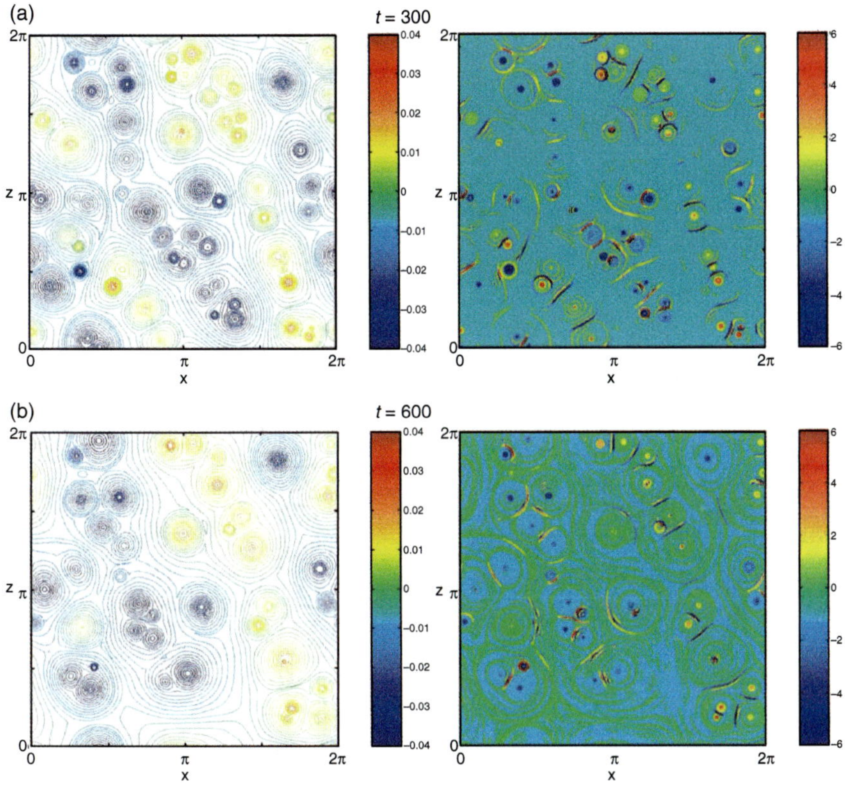

Figure 3.1. Views of interacting coherent structures resulting from a 2D MHD homogeneous turbulence simulation of a periodic square of $(2\pi)^2$ on a $(512)^2$ grid. Left panels are contours of magnetic flux at $t = 300$ and 600. Corresponding current density distributions are given in the right panels. For reference, the sound wave and Alfvén wave traveling times through a distance 2π are about 4.4 and 60, respectively. Colors indicate intensities in arbitrary units. Simulations with a strong uniform magnetic field B_0 will yield similar views in cross-sectional planes normal to B_0. (From Wu and Chang 2001.)

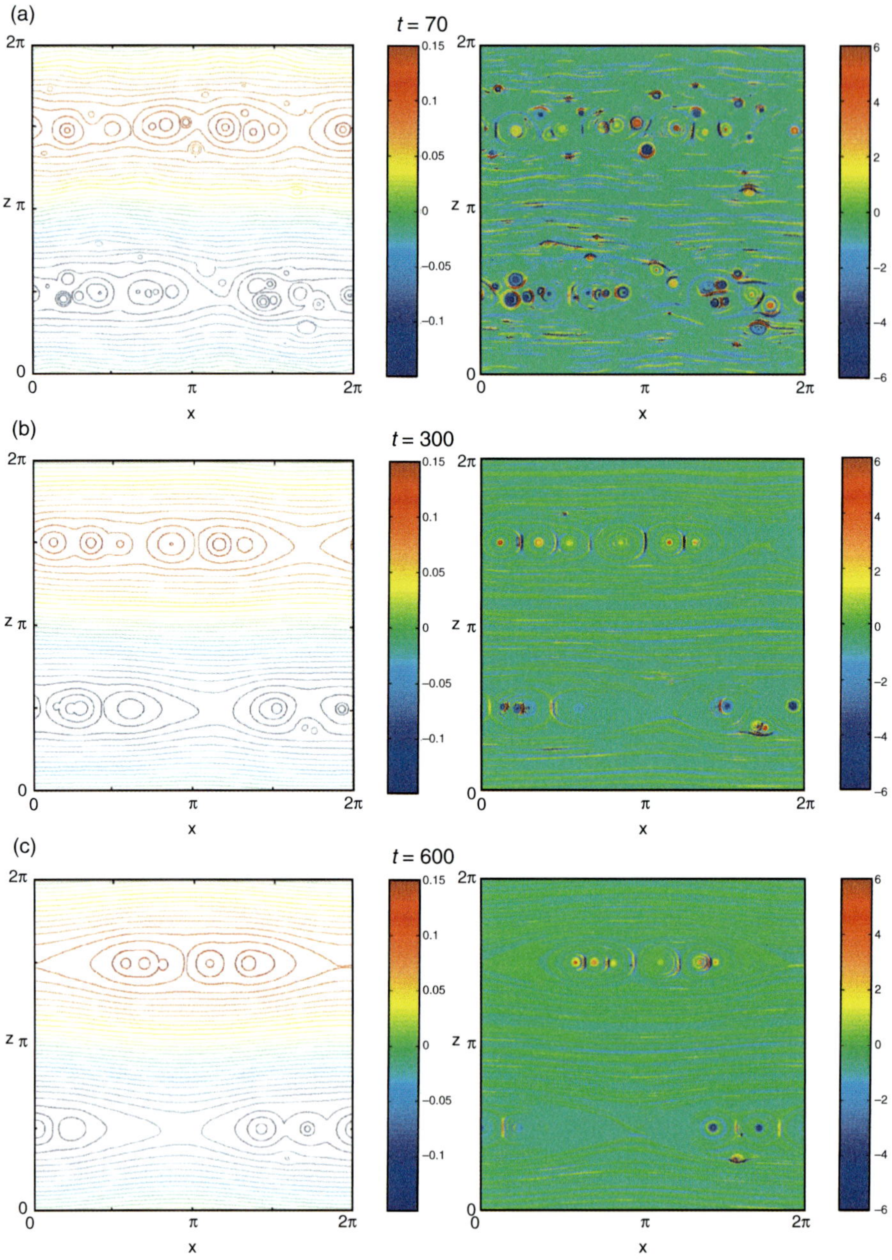

Figure 3.5. 2D MHD inhomogeneous turbulence simulations with magnetic shear of interacting coherent structures on a periodic square of $(2\pi)^2$ of a $(512)^2$ grid. Regions of strong magnetic shear (neutral sheet regions) are located at $z = \pi/2$ and $3\pi/2$. Left panels are contours of magnetic flux at $t = 70, 300, 600$. Corresponding current density distributions are given in the right panels. Colors indicate intensities in arbitrary units. For reference, the sound wave and Alfvén wave traveling times through a distance 2π are about 4.4 and 60, respectively. (From Wu and Chang 2001.)

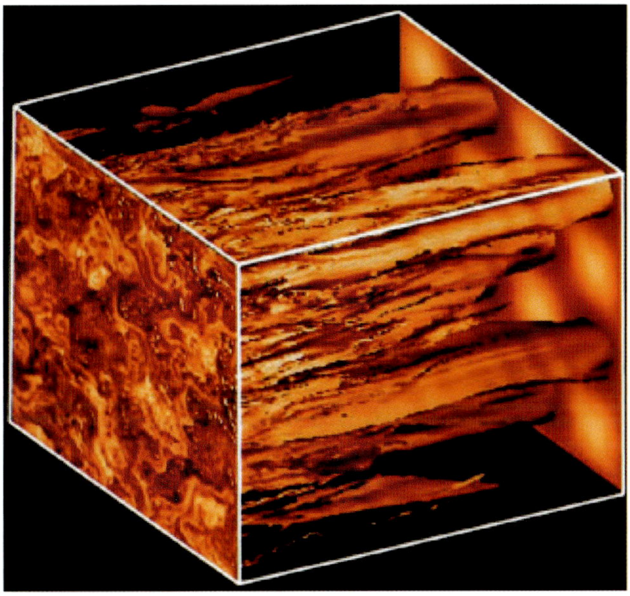

Figure 3.7. MHD simulation of a two-component model, with 80 percent of its energy in two-dimensional modes, exhibiting tangled flux tubes with considerable transverse complexity. (From Matthaeus et al. 2003.)

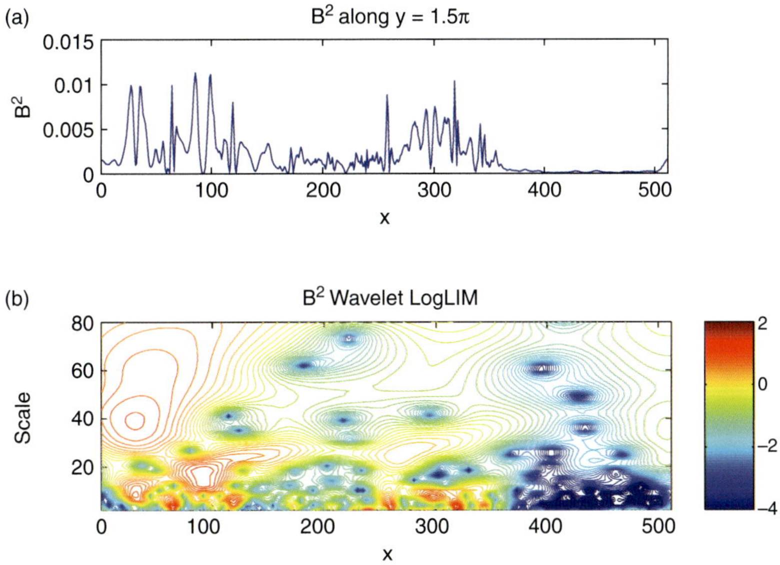

Figure 3.9. *LIM* of the fluctuating spatial series of the square of the magnetic field at the neutral sheet region obtained from a two-dimensional MHD simulation with sheared magnetic field geometry as described in Figure 3.5 using the Morlet transform. Colors indicate intensities in arbitrary units. (From Chang et al. 2004.)

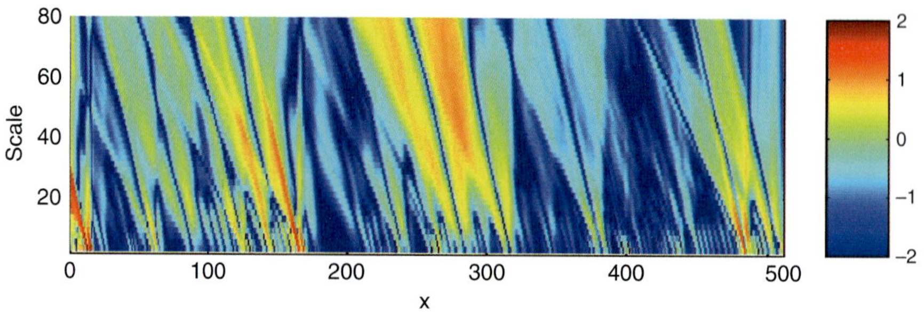

Figure 3.10. *LIM* of the current density J along the x-direction from a two-dimensional homogeneous MHD turbulence simulation as described in Figure 3.1 using the Haar transform. Colors indicate intensities in arbitrary units.

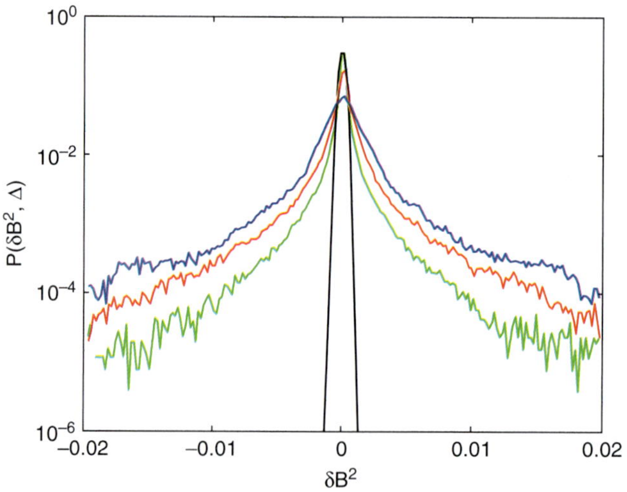

Figure 4.1. Probability distribution function $P(\delta B^2, \Delta)$ of a 2D MHD homogeneous turbulence simulation on a $(512)^2$ periodic lattice for $\Delta = 2(green), 4(red), 32(blue)$ units of grid spacing. A PDF for Gaussian fluctuations is also shown for comparison. (From Chang et al. 2004.)

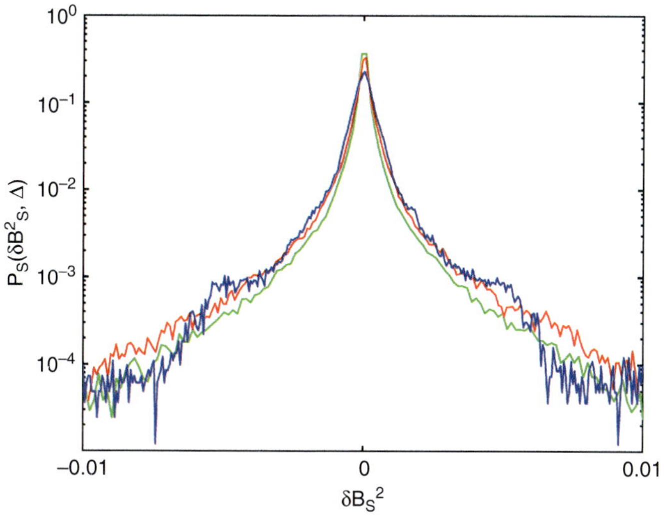

Figure 4.4. Scaled PDF for magnetic field energy density fluctuations δB^2 based on the one-parameter scaling form of Eq. (4.14) with $s = 0.335$ from a 2D MHD homogeneous turbulence simulation on a $(512)^2$ periodic lattice for $\Delta = 2$(*green*), 4(*red*), 32(*blue*) units of grid spacing. $\delta B_s^2 = \delta B^2 / \Delta^s$. (From Chang et al. 2004.)

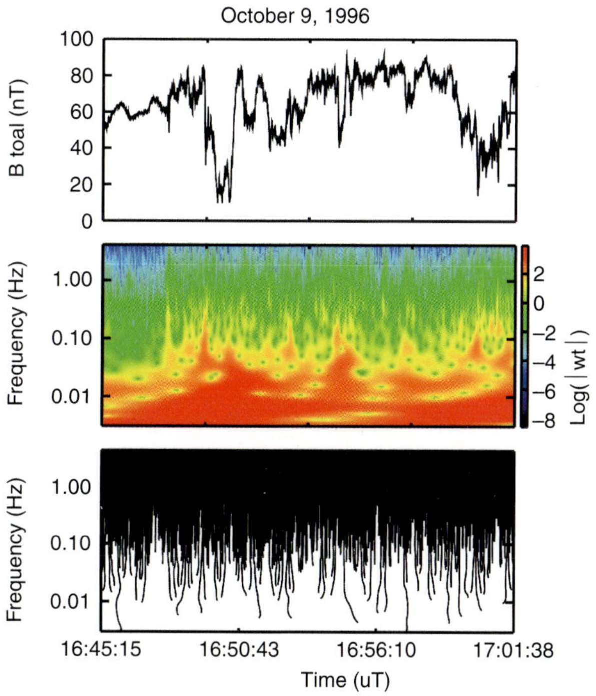

Figure 5.6. (Top panel): Segment of magnetic field data obtained by the Polar satellite on October 9, 1996 from 16:45:15 to 17:01:38 UT of the northern cusp region. (Middle panel): Its spectrogram. (Lower panel): Map showing the WTMM. (From Yordanova et al. 2004.)

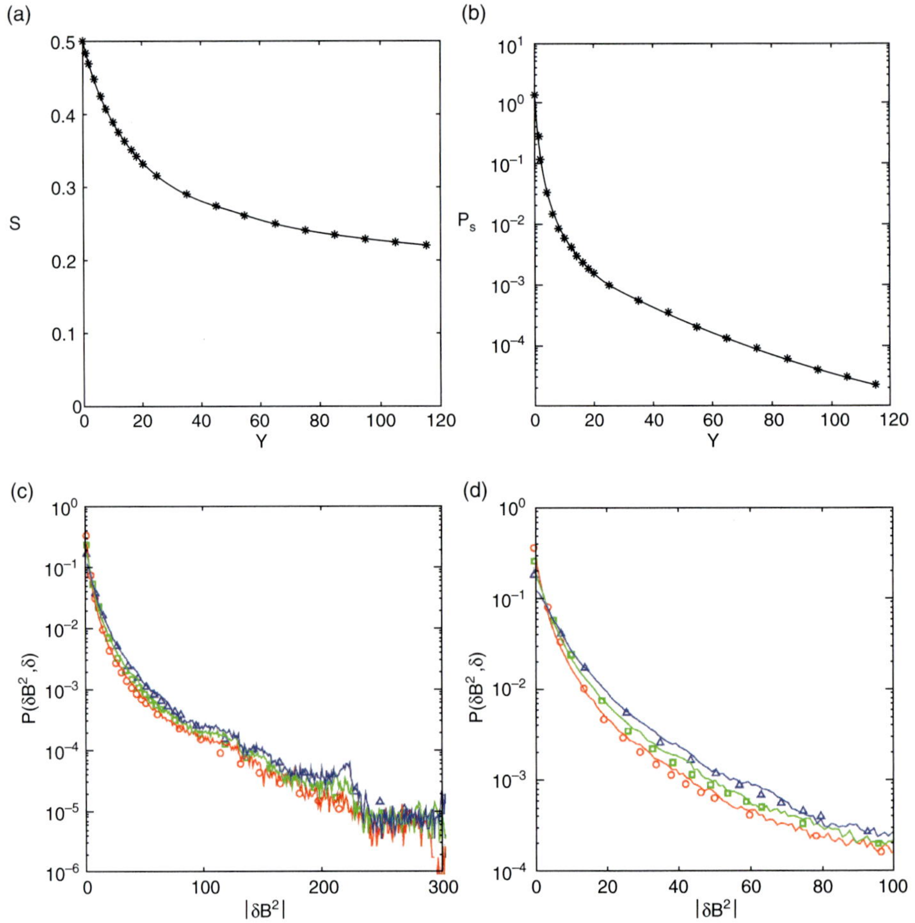

Figure 6.1. Fluctuations of B^2 based on a 2D homogeneous MHD simulation. (a) ROMA spectrum, $s(Y)$. (b) Scaled PDF, $P_s(Y)$. (c) Plots of computed PDFs from the ROMA relations (markers) and from the MHD simulation (solid curves). $\delta = 32\Delta$ (red circles), 48Δ (green squares): 96Δ (blue triangles). Δ (grid spacing) $= 2\pi / 1024$. $|\delta B^2|$ in units of $|\delta B^2_{\max}| / 800$. (d) Enlarged view of (c) for $|\delta B^2| \leq 100$. (From Wu and Chang 2011.)

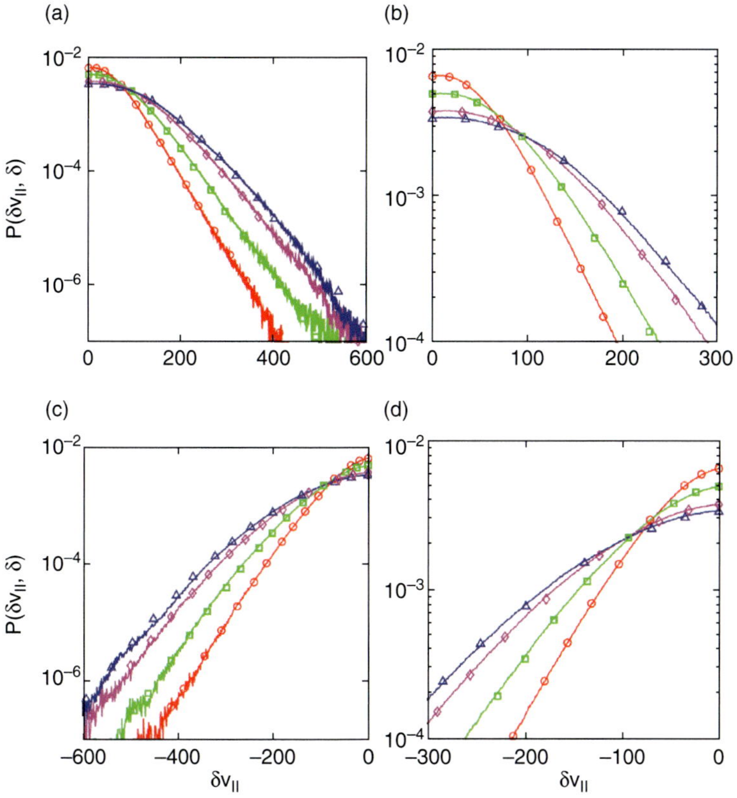

Figure 6.4. Plots of $P(\delta v_{\|},\delta)$ for fully developed isotropic fluid turbulence in the inertial range as discussed in Figures 6.2 and 6.3: solid curves are from the simulation data and markers are from ROMA scaling relations: red (circles) for $\delta = 24\Delta$; green (squares) for $\delta = 48\Delta$; magenta (diamonds) for $\delta = 96\Delta$; blue (triangles) for $\delta = 128\Delta$. (a) and (b) are for positive $\delta v_{\|}$; (c) and (d) are for negative $\delta v_{\|}$. (b) and (d) are enlarged plots. (From Wu and Chang 2011.)

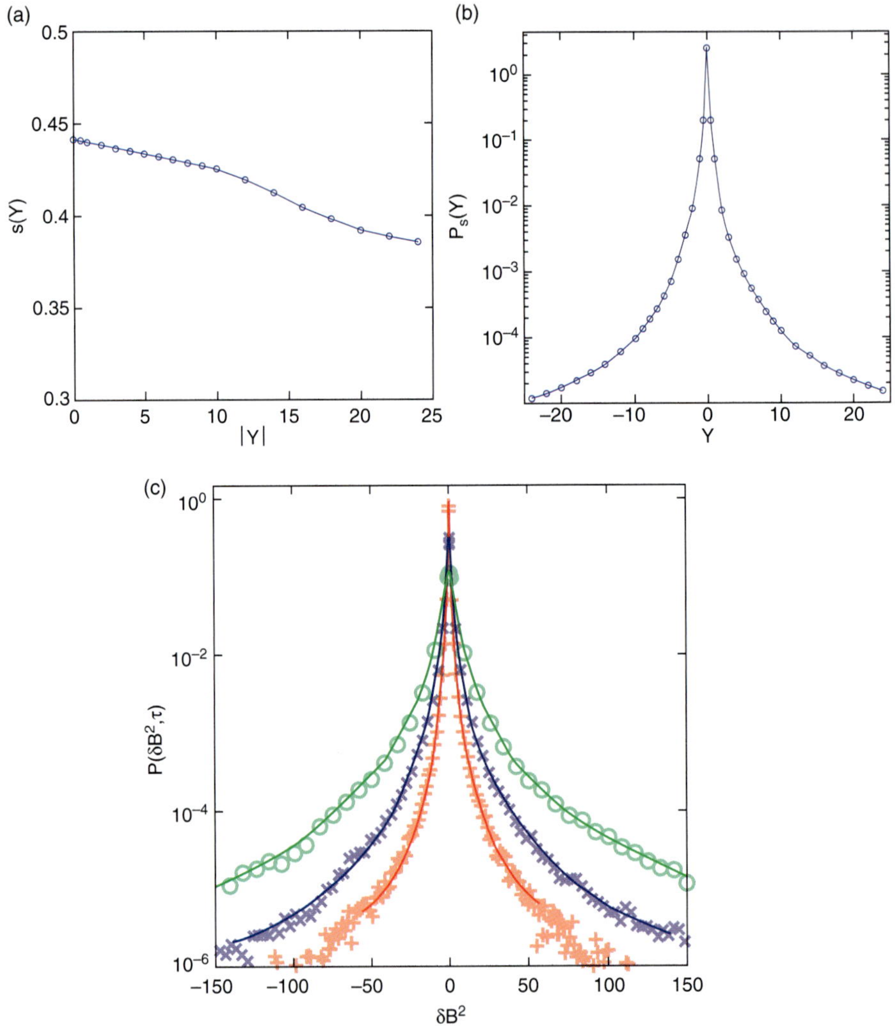

Figure 6.5. Fluctuations of B^2 based on the solar wind data. (a) and (b) ROMA spectrum $s(Y)$ and scaled PDF $P_s(Y)$. (c) Computed PDFs $P(\delta B^2, \tau)$ from ROMA relations (solid curves) and from solar wind observations (markers). δB^2 in units of $(nT)^2$. $\tau = 1000s$ (green o), $96s$ (blue x). $9s$ (red +). (From Wu and Chang 2011.)

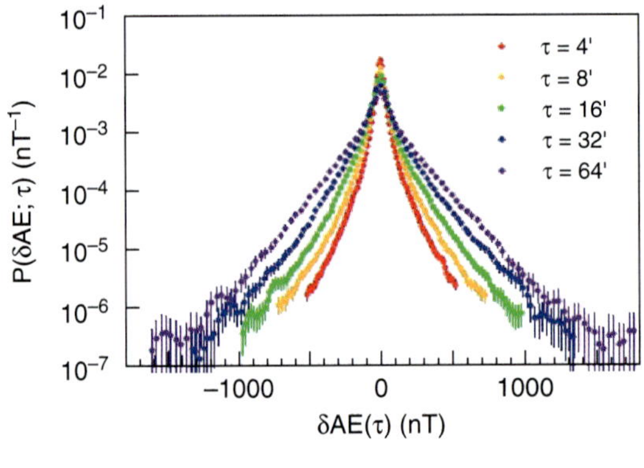

Figure 6.6. PDFs $P(\delta AE;\tau)$ for AE increments $\delta AE(\tau)$ at $\tau = 4,8,...,64$ minutes. (From Consolini and De Michelis 2011.)

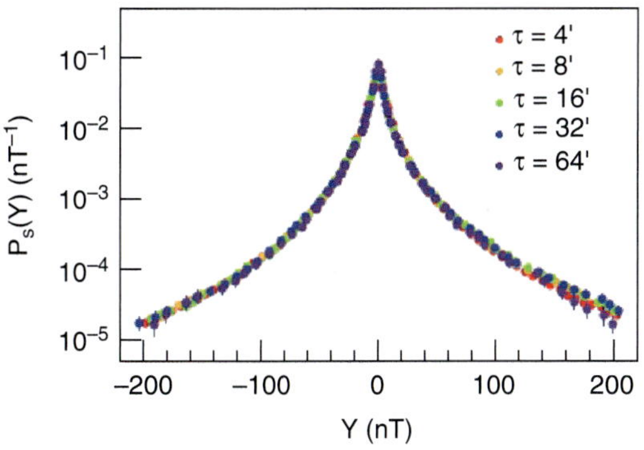

Figure 6.8. Scaled PDF $P_s(Y)$ for the probabilities of the AE increments of Figure 6.6 in units of $(nT)^{-1}$. Y is in units of nT. (From Consolini and De Michelis 2011.)

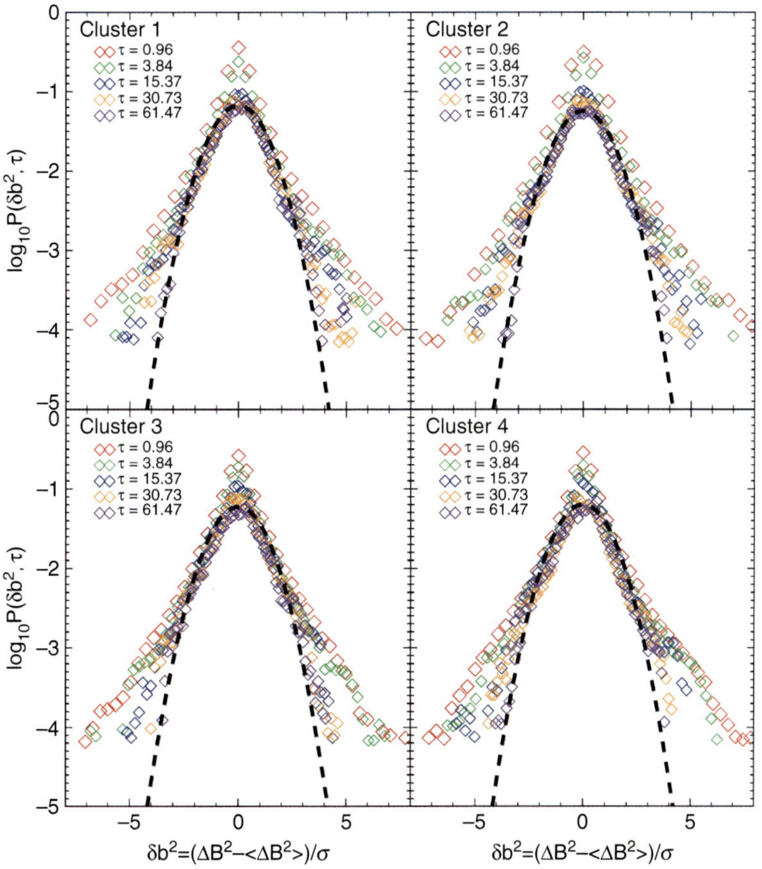

Figure 6.10. Typical PDFs of the magnetic energy density fluctuations measured by the four Cluster satellites for the cusp passage described in Figure 6.9; the PDFs have been scaled with respect to their variance. The scales of τ are color coded and given in seconds. (From Echim et al. 2007.)

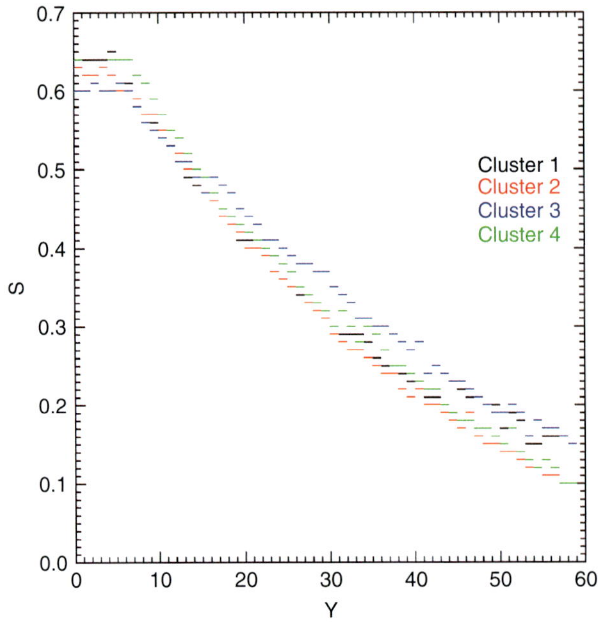

Figure 6.11. Rank-ordered multifractal spectra for the Cluster passage of Figures 6.9 and 6.10. (From Chang et al., 2010; Echim and Lamy, private communication, 2010.)

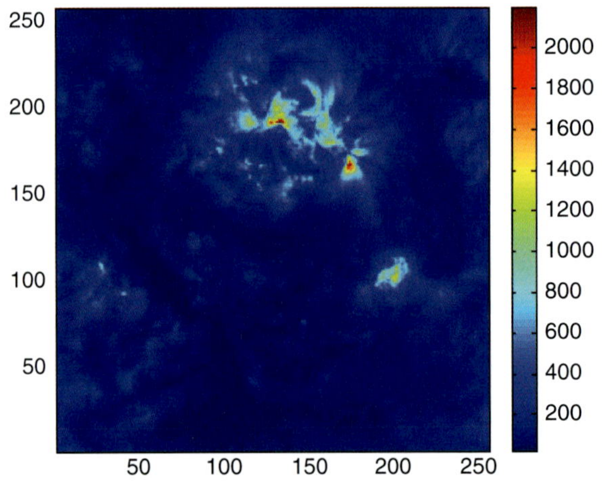

Figure 6.12. The Luminosity $w(\mathbf{r})$ in arbitrary units of a SOHO image. Only the central portion of the Sun disk with the linear dimensions 1040×1040 Mm (256×256 pixels with 5.6 arcsec resolution) was studied. (From Wu et al. 2011.)

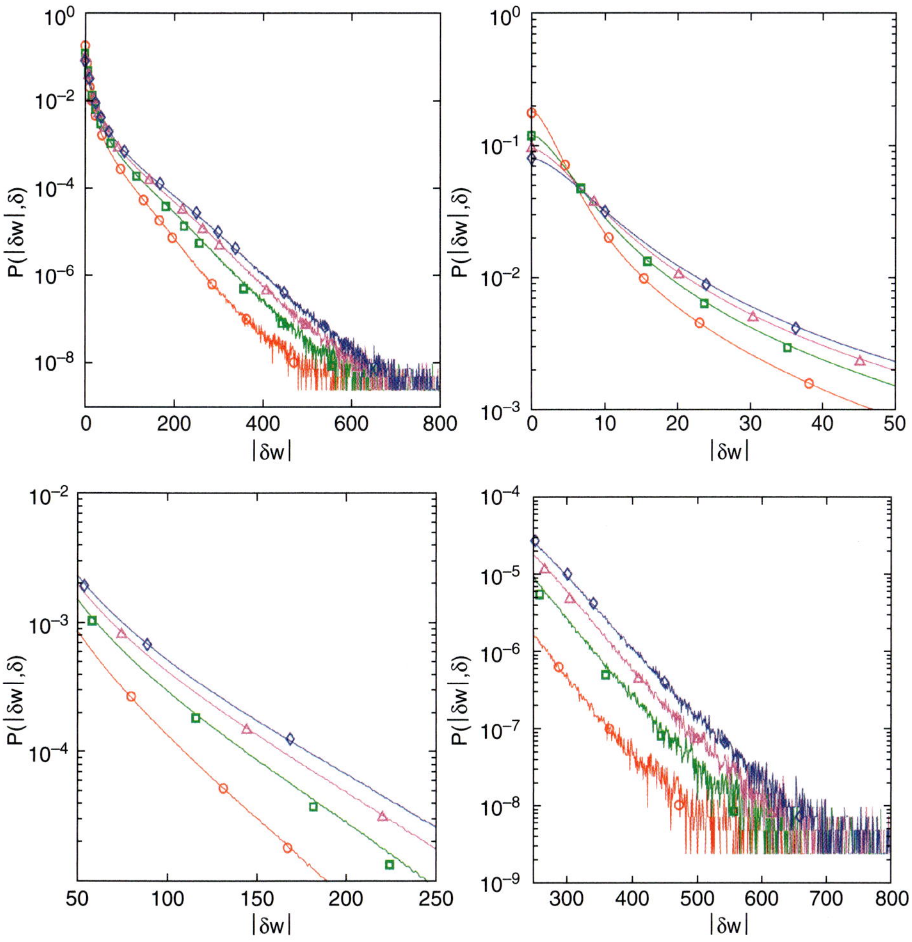

Figure 6.15. Computed PDFs $P(\delta\left|w(r)\right|,\delta)$ from ROMA relations (solid curves) and from SOHO observations (markers). $\delta = 2$ (red circles), 4 (green squares), 6 (magenta triangles), 8 (blue diamonds). (From Wu et al. 2011.)

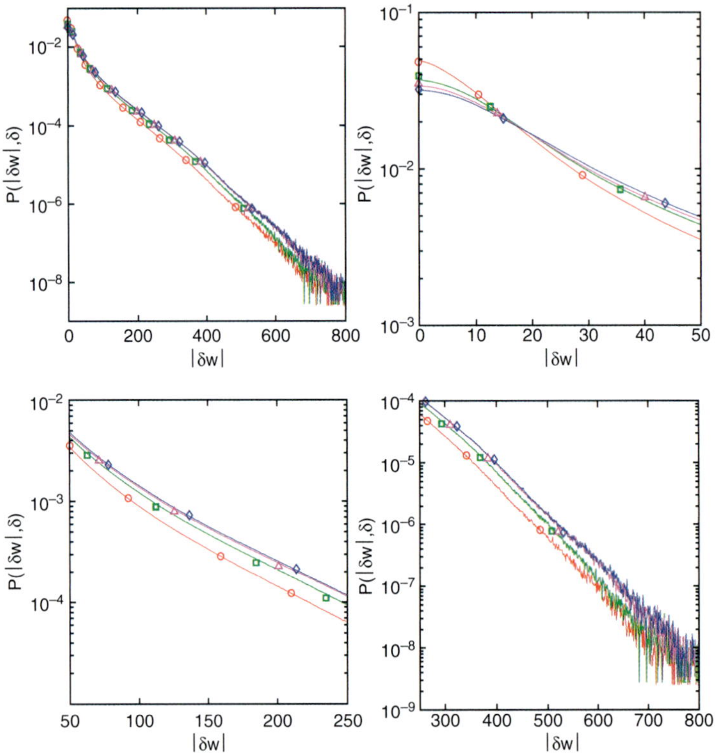

Figure 6.16. Computed PDFs $P(\delta|w(r)|,\delta)$ from ROMA relations (solid curves) and from SOHO observations (markers). δ = 20 (red circles), 40 (green squares), 60 (magenta triangles), 80 (blue diamonds). (From Wu et al. 2011.)

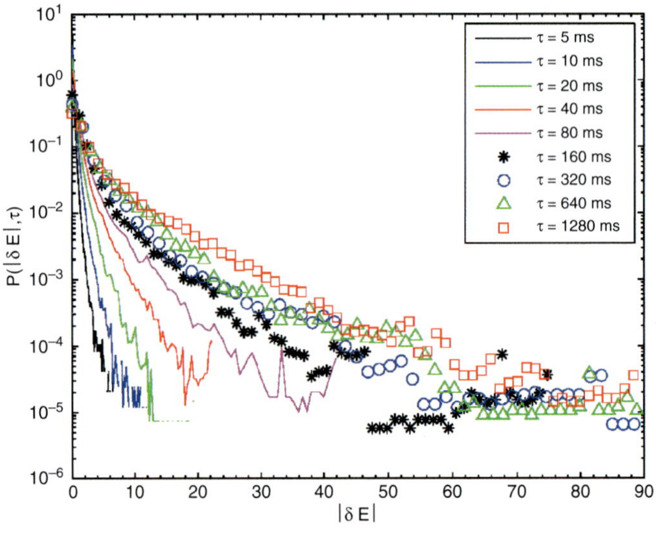

Figure 6.18. PDFs of the BBELF electric field fluctuations $|\delta E|$ in units of mV/m are non-Gaussian and strongly scale-dependent, bearing the typical intermittent characteristics. (From Tam et al. 2010.)

1

Introduction

1.1 What Is Complexity?

Complexity can arise from an interacting system composed of many individual elements. If the rules of engagement among the interacting elements are nonlinear and long-ranged, quite often large-scale structures are generated – each structure being composed of many individual elements behaving more or less coherently together. Such "coherent structures" can have varied sizes. In the stock market, examples of the coherent structures are the hedge funds, investment banks, pension funds, mutual funds, wealthy individuals, investment clubs, and so forth. And the ensuing behavior resulting from the diversified interactions of these structures may become very complicated and sometimes almost incomprehensible. Similar situations may occur in dynamical plasmas (Chang 1999). When large numbers of electrons and ions interact nonlinearly in an electromagnetic environment, coherent structures of assorted types and sizes are generated. These coherent structures may be in the form of double layers, ion holes, electron holes, drift Alfvén vortices, flux tubes, current filaments, stationary and convecting structures, and so forth. When such (and partially formed) structures interact, the ensuing stochastic behavior can become very complex and generally not easily deducible from the basic equations that characterize the nonlinear interactions of the original elements of the underlying dynamical plasma – even when analyzed with modern-day, large-scale supercomputers.

We shall consider in this monograph selected examples of complexity phenomena related to the observed plasma processes in the space environment, primarily those obtained from the heliosphere – a bubble created by the capsular radiation originating from the sun – such as those pertaining to the solar flares, the solar wind, and the Earth's magnetosphere and ionosphere.

To provide a historical introduction as well as a conceptual overture to the subject of complexity, we describe in the following section some basic ideas pertaining to the phenomenon of criticality in thermodynamic equilibrium.

1.2 Criticality

A classical example of complexity is the phenomenon of criticality, which received its initial attention during the late 1960s and early 1970s (Stanley 1971). Consider, for example, the onset of "critical opalescence" in binary fluids. This phenomenon may be easily demonstrated in a lecture hall using a test tube containing, for example, a mixture of hexane and nitrobenzene. At room pressure and temperature, such a mixture when appropriately proportioned may separate into two stable coexisting liquid phases in thermodynamic equilibrium; one being hexane-rich, which is lighter, and the other being nitrobenzene-rich and heavier. The separation between these two individual liquid phases in the test tube is characterized by a distinct horizontal meniscus. As one holds the test tube in the hand while lecturing, the mixture warms up and eventually becomes one single homogeneous mixture beyond some elevated "critical" temperature. As this transition takes place, the mixture becomes milky white so that a light beam shining on it cannot pass through and instead will be scattered strongly creating the illusion that the entire mixture glows – "critical opalescence." If the lecture room is darkened and a laser beam is shined upon the mixture, the laser beam will be scattered randomly producing dots of light dancing all over the walls of the lecture hall. When a blue laser beam is employed, blue dots will be scattered in each and every direction; red laser beam, red dots; green laser beam, green dots; and so forth, indicating that the mixture is composed of a collection of heterogeneous blobs of particles with linear dimensions covering all wavelengths of the visible light. Apparently, at the critical temperature, the molecules of the liquid begin to interact over very long distances and thereby generating blobs of molecules that correlate strongly such that each blob moves essentially coherently in unison. The linear dimension of each blob can in fact become as large as that of the wavelength of light and thus capable of reflecting the light waves. Because most molecules have linear dimensions on the order of angstroms and the wavelengths of the visible light are of the order of thousands of angstroms, the blobs generally can contain on the order 10^9 molecules or more. As the blobs wander around randomly in the mixture, they scatter light waves incoherently in all directions. The fact that the mixture looks milky white indicates that the dimensions of the blobs can be of the extents of all wavelengths of the visible light, and it is the same reason why milk is milky white. In fact, the blobs may have linear dimensions extending beyond the wavelengths of the visible spectrum of light in both directions.

After a little while, though, as the temperature of the mixture continues to rise, light begins to pass through the mixture without being strongly scattered. The

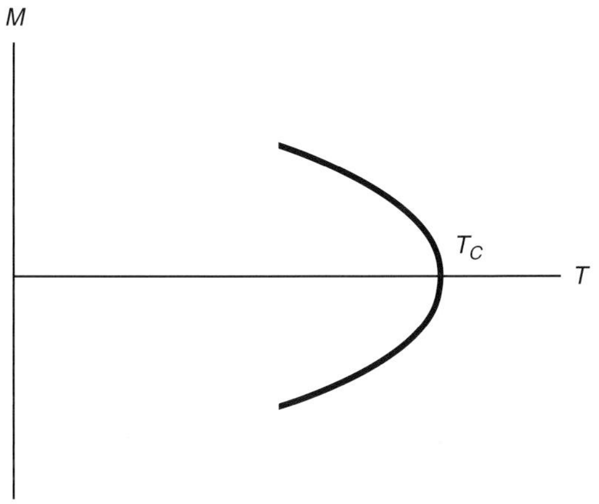

Figure 1.1. Phase Diagram of Magnetization (M) versus Temperature (T) for a Ferromagnet near the Curie Temperature (T_c).

mixture now has become a single phase of a transparent fluid medium and the molecules no longer correlate over long distances.

At the critical temperature – the temperature at which the mixture exhibits critical opalescence, experimentalists have found that such a binary mixture exhibits extraordinary thermodynamic properties. To start with, the light scattering by the binary mixture becomes very strong and erratic. And, the specific heat, for example, may diverge weakly above the critical temperature as it is approached. There are also many other peculiar thermodynamic properties of the mixture near the critical temperature, that we could but shall not enumerate here.

To gain some feeling of such extraordinary behavior for thermodynamic systems near criticality, let us consider an even simpler example of equilibrium phenomena, the behavior of an ordinary room-temperature magnet (ferromagnet). If one dips such a magnet in a box of steel thumbtacks and then lifts the magnet upward, a number of thumbtacks will generally be attached to the magnet and be lifted together with the magnet. When one suspends the magnet in a laboratory oven, where it may be seen through a glass window, and turns up the temperature, T, of the oven gradually, one observes that the attached thumbtacks begin to fall away one by one from the magnet. At some specific elevated critical temperature, suddenly all of the remaining thumbtacks fall away together. The number of thumbtacks attached to the magnet is a measure of the so-called magnetization, M, of the magnet. As the temperature of the magnet is increased, the magnetization of the magnet decreases. At and above some specific critical temperature, T_c, the magnet is no longer magnetized. This is the so-called Curie temperature of the magnet.

Let us visualize the magnet as a three-dimensional lattice composed of little magnets such that each can either point in the upward or downward direction with a discrete spin value of +1 or –1 – magnetic dipole moments of atomic spins or Ising spins. Thus there is an up-down symmetry associated with the magnetization for such a magnet. The phase diagram M versus T near the critical temperature will then have a shape such as that depicted in Figure 1.1.

From the basic theory of statistical thermodynamics, we learned that all thermodynamic properties at equilibrium are characterized by the partition function. For the magnet, the partition function Z is,

$$Z \equiv \sum_{a.p.c.} \exp\{-\mathcal{H}(s)/kT\} \tag{1.1}$$

where k is the Boltzmann constant and $\mathcal{H}(s)$ is the Hamiltonian of the spin system for a particular configuration of the Ising spins on the lattice, and, for which we shall include only the nearest neighbor interactions of the spins. Thus we write

$$\mathcal{H}(s) = -J \sum_{<i,j>} s_i s_j \tag{1.2}$$

where J is the interaction energy between the adjacent spins and $<i,j>$ are nearest neighbor sites on the lattice. Each term $\exp\{-\mathcal{H}(s)/kT\}$ in the sum of Eq. (1.1) is proportional to the probability for the system to be in a particular configuration of the spins on the three-dimensional lattice. The summation sign associated with the definition of the partition function represents the sum over *all possible configurations* (a.p.c.) of the up and down spins on the lattice. Thus the meaning of the "partition function" is the "sum of all probable states" and is the normalization constant for the probabilities. The preceding formulation may be generalized to include an external magnetic field h by including a term $-h\sum_i s_i$ on the right-hand side of Eq. (1.2).

Also, from equilibrium statistical thermodynamics, we know that the magnetization M is given by

$$M(T,h) = -(\partial F/\partial h)_T \quad \text{with} \quad F(T,h) = -kT \ln Z \tag{1.3}$$

where $F(T,h)$ is the (Gibbs) free energy and the subscript T emphasizes the fact that the partial differentiation is taken with the temperature held constant. Because each term in the sum of the partition function is an analytic function, any finite sum of the terms in the partition function must also be analytic. Thus one "concludes" that all thermodynamic properties of the magnet would have to be analytic. Applying this result to the phase diagram M-T of Figure 1.1, one "realizes" that the shape of the diagram must be in the form of a symmetric parabola near the critical or "Curie" temperature, T_c. Therefore, in a log-log plot of $|M|$ versus $T_c - T$ near and below the critical temperature, the shape of the phase diagram near the critical temperature would be a straight line with a slope equal to $-1/2$. At least this is what the theoreticians used to believe. Experimentalists, however, discovered that while the log-log plot of such a phase diagram was indeed approximately a straight line, the slope was nevertheless closer to the value of $-1/3$ instead of $-1/2$. And, in fact, the value of the slope for real magnets appeared to be an irrational number within the experimental limits (Heller and Benedek 1962). This dilemma between "theory" and "observations" could not be resolved even with very precise measurements close to the critical temperature. At this point, the theoreticians were forced to reconsider the logic that led them to the conclusion that the slope would be $-1/2$. They eventually realized that the only way for the magnet to behave nonanalytically at the critical temperature was if the thermodynamics of the system at criticality involved the full degrees of freedom of the system and therefore must be described by the entirety of the sum of the partition function. In other words, all degrees of freedom of the magnet must be accounted for in the thermodynamic description at criticality – a phenomenon of true infinity. The rationalization here is that near criticality, the spins correlate with each other so strongly that they form coherent domains of varied sizes, each being much larger than the size of the individual spins. And the statistical properties of the magnet near criticality must include all the degrees of freedom involving the interactions of the coherent spin domains of all sizes and shapes.

Experimentalists were able to demonstrate this by looking at the scattering of the neutrons by the magnet near criticality. Indeed, the scattering becomes abnormally large and random. Furthermore, neutrons of any energy were found to be scattered this way indicating that the anomalous scattering was produced by the random mixing of coherent domains of spins of varied sizes. This abnormal behavior manifests in many thermodynamic characteristics of the magnet near criticality. Observations indicate that most of such singular behavior may be characterized in leading order by power law singularities. Associated with such singularities one speaks of the critical exponents. For example, related to the zero-field magnetization M and isothermal susceptibility (i.e., the response of M due to the application of h)$\chi_T = (\partial M / \partial h)_T$, it is conventional to write the asymptotic critical behavior near the Curie point T_c as

$$M \sim t^{\beta} \text{ and } \chi_T \sim t^{-\gamma} \tag{1.4}$$

where $t \equiv |T - T_c| / T_c$ is the so-called reduced temperature, and (β, γ) are the corresponding critical exponents. For thermodynamic quantities such as the susceptibility, specific heat, and so forth, the critical exponents may be defined by limits approaching the critical temperature from both above and below. While experimentally the exponents obtained from the two limits may be different, theoretical arguments generally suggest that they are the same. We shall not dwell on these differences in the present discussion.

There is a set of alphabets of critical point exponents characterizing the leading singular behavior of different thermodynamic quantities of the ferromagnet. For example, the asymptotic behavior of the specific heat $C_h = -T(\partial^2 F / \partial T^2)_h$ at constant magnetic field h as a function of temperature, and the magnetization M at $T = T_c$ as a function of the external magnetic field h are characterized by

$$C_h \sim t^{-\alpha} \text{ and } M \sim h^{-\delta} \tag{1.5}$$

For the ferromagnetic system to exhibit criticality, the correlations among the spins are very long-ranged and there exist many correlation scales (as characterized by the anomalous scattering of the neutrons with various energies). Therefore it is reasonable to expect the system to exhibit some sort of invariance property under coarse-graining scale transformations. The mathematical consequence of this invariant property (based on the renormalization group theory to be discussed later in this monograph) is that, close to criticality, most parameters and thermodynamic properties would scale linearly with the coarse-graining transformations, for example,

$$t \to t' = e^{\lambda_t \ell} t, \quad h \to h' = e^{\lambda_h \ell} h, \quad F \to F' = e^{\lambda_F \ell} F \tag{1.6}$$

where the constants $(\lambda_t, \lambda_h, \lambda_F)$ are scaling powers and ℓ is the scaling parameter. The consequence of this linear transformation property is that the parameters and thermodynamic quantities scale as power laws such as those given in Eqs. (1.4) and (1.5). Now, if the functional relation between F and (t,h) would remain the same under scale transformations, then F would have the form of a generalized homogeneous function near criticality:

$$F(e^{\lambda_t \ell} t, e^{\lambda_h \ell} h) = e^{\lambda_F \ell} F(t,h) \tag{1.7}$$

Before the advent of the renormalization group theory introduced by Wilson (1971), these ideas were heuristically hypothesized by Kadanoff (1966) and others, and they had been collectively called the scaling theory or scaling hypothesis. (For a brief review, see Stanley [1971].) Equation (1.7) may be recast in the more commonly adopted scaling form as

$$F(t.h) = t^{2-\alpha} F(h / t^{\beta\delta}) \tag{1.8}$$

where $2-\alpha = \lambda_F / \lambda_t$ and $\beta\delta = \lambda_h / \lambda_t$. We note that the scaling powers in expressions such as Eq. (1.7) are determined only up to a common scale factor. The generalized homogeneous form for F also implies the following exponent equalities.

$$\alpha + 2\beta + \gamma = 2 \quad \text{and} \quad \beta(\delta + 1) = 2 - \alpha \tag{1.9}$$

which were proven to hold generally as inequalities (left-hand sides greater than or equal to right-hand sides) by Rushbrooke (1963) and Griffiths (1965), respectively.

In addition to the pure thermodynamic singularities such as those characterized by Eqs. (1.4) and (1.5), the spin-spin correlation function Γ of a ferromagnet also exhibits power law behavior at criticality. The correlation length, ξ, diverges as the critical temperature is approached, at which Γ diverges as a power of the distance between spins, $r = |\mathbf{r}|$ (Fisher 1964):

$$\xi \sim t^{-\nu}, \quad \Gamma \sim r^{2-d-\eta} \tag{1.10}$$

where ν is the correlation length exponent, η is called the correlation function exponent and d is the spatial dimension of the spin lattice representing the ferromagnet.

In a similar manner as we have argued with respect to the free energy, we may use the invariant concepts to arrive at the scaling hypothesis for the correlation function at criticality as a generalized homogeneous function (which again may be supported more rigorously by the renormalization-group theory):

$$\Gamma(e^{\lambda_r \ell} r, e^{\lambda_t \ell} t, e^{\lambda_h \ell} h) = e^{(d-2+\eta)\ell}\Gamma(r,t,h) \tag{1.11}$$

These correlation function and correlation length exponents are not independent of the thermodynamic exponents discussed earlier. Because the isothermal susceptibility is given by

$$\chi_T = \int d^d r\, \Gamma^{-1}(r,t,h) \tag{1.12}$$

we have without further assumption $\gamma = v(2-\eta)$. The choice of a fixed value $\lambda_r = -1$ for the scaling power of distance fixes the scale factor of the remaining powers λ_h and λ_t; for example, $v = 1/\lambda_t$. If it is further assumed that the scaling power of the free energy $\lambda_F = d$ (so that the free energy in a correlation volume, $F\xi^d$, is finite in the vicinity of the critical point), then the hyperscaling relationship (Fisher 1969, 1973) holds: $dv = 2-\alpha$, which was derived by Josephson (1967) as an inequality.

Finite Size Scaling

Observational, experimental, or numerically simulated systems near criticality are of finite sizes. For a system characterized, for example, by the finite linear dimension L, Fisher (1971) suggested that the free energy near criticality scale with the reduced temperature t as follows:

$$F(t,L) = L^{-\psi} F_L(tL^{1/v}) = \xi^{-d} F_s(L/\xi) \tag{1.13}$$

where $\psi = d = (2-\alpha)/v$ and $\xi \sim t^{-v}$ is the correlation length of the infinite system. Expression (1.13) indicates that for a finite size system near criticality, the observational, experimental, or simulated data would collapse to a single curve characterized by F_L.

We may easily arrive at the finite size scaling (FSS) hypothesis (1.13) by postulating that $F(t,L)$ scales as a generalized homogeneous function as follows:

$$F(e^{\lambda_t \ell} t, e^{\lambda_L \ell} L) = e^{\lambda_F \ell} F(t,L) \tag{1.14}$$

Because F is an extensive variable that scales with the size of the system and because of the hyperscaling relation, we have $\lambda_F / \lambda_L = -d = -(2-\alpha)/v$. With $\lambda_t / \lambda_L = -1/v$, we have immediately Eq. (1.13).

The preceding discussions have been couched in terms of the ferromagnetic critical behavior; however, analogous expressions hold true for a variety of systems near their critical points. For example, the critical behavior of the binary mixture discussed earlier can be mapped onto that of the ferromagnet. Criticality can be divided into universality classes that share the same critical point exponents (with a judicious choice of scaling variables) and the same scaling function descriptions. The specification of universality classes generally depends (primarily) on the symmetry of the order parameter or parameters (such as M in the case of the ferromagnet) and the spatial dimension d of the system.

1.3 Self-Organized Criticality

In Section 1.2, we discussed the phenomenon of criticality with examples from equilibrium thermodynamic systems. In space plasmas, we are mainly concerned with complexity phenomena that arise within nonequilibrium systems far from equilibrium. In 1987, Bak and colleagues suggested the possibility for slowly driven dissipative systems (generally far from equilibrium) to evolve naturally toward a dynamical criticality with properties similar to the critical behavior for equilibrium thermodynamic systems and constructed ideal models (in particular, sandpiles) to illustrate the concept. In their sandpile model, if the local slope of the sandpile was perturbed (e.g., by the addition of one grain of sand locally) such that it becomes larger than a critical value, the pile would collapse locally, which might then cause their adjacent slopes to exceed the critical value and hence inducing new localized collapses. This process would continue until the sandpile becomes stable. And this "avalanching process" might be induced again and again at other localities until the average slope of the sandpile became barely stable with respect to small perturbations. At that moment, the statistical properties of the sandpile system would exhibit similar scale-invariant characteristics as those for critical points in equilibrium thermodynamic systems. Bak and colleagues (1987) suggested that such organization to criticality would generally take place for dynamical systems far from equilibrium without specific tuning. Thus it would be different from the classical equilibrium thermodynamic critical phenomena, which require careful tuning of the parameters such as the temperature of a system. They dubbed this type of phenomenon as "self-organized criticality," or SOC.

It is now understood that some tuning is still needed for SOC and the low frequency $1/f$ flicker noise behavior of Section 2.2 to be discussed in the following text, which Bak and colleagues (1987) predicted, does not really occur for the original model that they suggested. Thus modified sandpile models and other cellular automaton or coupled map lattice models such as the forest fire, Faraday induction models, and so forth have sprouted up within the past twenty-five years to address analogous and/or more generalized types of processes. We shall return and address such phenomena in the context of scale invariants for model dynamical systems mimicking real space plasma processes in Chapter 2.

1.4 Fluctuations, Intermittency, and Multifractals

Another common characteristic feature of complexity is the phenomenon of intermittency. We recall complexity is generated by the complex stochastic interactions of nonlinearly generated coherent structures of varied sizes in an infinite degrees of freedom system. Probability distribution functions (PDFs) of the fluctuations of such turbulent states are commonly non-Gaussian and their forms may vary from scale to scale. See, for example, Figure 4.1. When this happens, it is a strong indication that the fluctuations are not self-similar under scale transformations. And such fluctuations are generally intermittent as discussed in the following text.

1.4.1 Wavelet Transform

One way to visualize the phenomenon of intermittency is to consider the wavelet transform of such turbulent fluctuations. Traditional procedures for analyzing fluctuations are to perform Fourier transforms of the observational, numerical, or experimental data. The basis functions of the Fourier transform are sines and cosines. Each such basis function extends in space (or time) all the way from minus infinity to plus infinity. Thus Fourier transforms essentially wipe out the localized information of the fluctuations resulting from the confined interactions of the coherent structures. Therefore, for turbulence studies, it will be more appropriate to consider transforms whose basis functions are localized and scale dependent instead of sines and cosines. This type of transforms has the generic name of "wavelet transforms" (Farge 1992).

To be more specific and restricting the discussion to one dimension, a wavelet transform replaces (transforms) a fluctuating spatial (or temporal) series into the coefficients of a set of basis functions that are localized in space (or time) with different scales in space (or time). The basis functions are usually chosen such that they are square-integrable and generated by a single mother wavelet – a rescalable and translatable function $\psi[(x-b)/a]$ or $\psi[(t-b)/a]$ where a is the scale and b defines the translation along x or t.

To obtain the coefficients of the wavelet transform for a particular fluctuating spatial (or temporal) series $S(x)$, we evaluate the convolution integral as follows:

$$C(a,b) = (K/a)\int S(x')\,\psi[(x'-b)/a]dx' \tag{1.15}$$

or its time domain equivalent, where K is a normalization constant. From $C(a,b)$, we can calculate, for example, the normalized power NP(a,b):

$$\mathrm{NP}(a,b) = \left|C(a,b)\right|^2 / \left\langle \left|C(a,b)^2\right| \right\rangle_x \tag{1.16}$$

where $<\cdots>_x$ denotes the spatial (or temporal) average; and, here, the average is essentially the Fourier-transformed power for the entirety of the fluctuations. The normalized power is a power spectrum of both scale and location; thus, it is sometimes called the Local Intermittency Measure, or *LIM*. By definition, *LIM* is equal to one for the Fourier transform. In usual applications, only a truncated finite set of coefficients are used to perform practical calculations. Figure 1.2 is a color representation of a sample *LIM* of the magnetic energy density fluctuations obtained

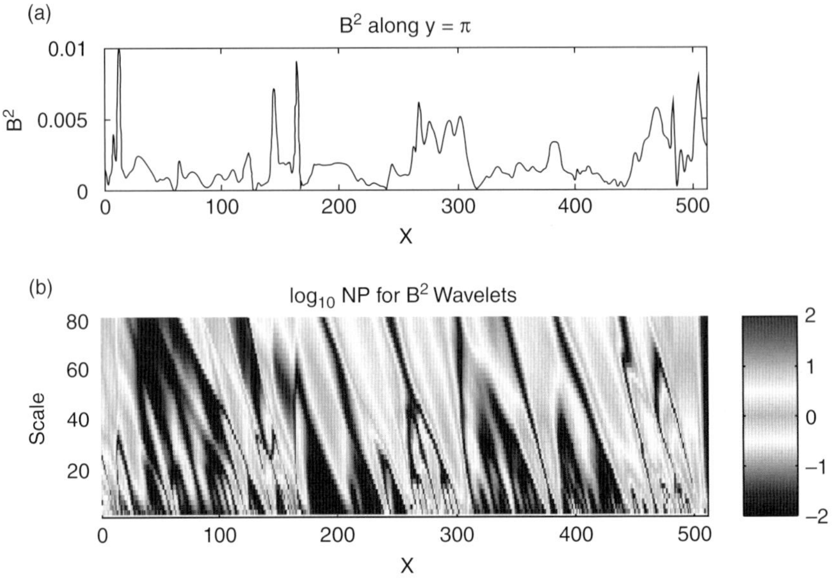

Figure 1.2. Local Intermittency Measure (*LIM*) or Normalized Power (*NP*) of the fluctuations of the magnetic energy density B^2 in arbitrary units at constant values of y and t obtained from a 2D (512×512) MHD homogeneous turbulence simulation using the Haar transform. The x-axis and scales are in units of grid spacing. (For a color version of this figure, please see the color-plate section.)

from a 2D magnetohydrodynamic (MHD) simulation using the Haar transform as defined in Chapter 4. The scale-dependent, sporadic, and localized powers of the fluctuations are vividly displayed. This sample *LIM* plot exhibits the intermixing geometrical patterns that are not always commensurate with each other. The occurrence of incommensurate patterns is a good indicator of intermittent behavior among observed fluctuations. We note that the intermittency generally increases as the scale is reduced. And the intensities can be rather strong at small scales – even when the fluctuations are quite intermittent and localized.

1.4.2 Geometrical Interpretation of Intermittency

To provide a physical and geometrical feeling for intermittency such as those characterized in Figure 1.2 by wavelet transforms, we consider the concept of fractals. In ordinary Euclidean geometry, the area of a two-dimensional square may be expressed as the square of the length of one of its sides, and the volume of a three-dimensional cube as the cube of the length of one of its sides. Generalizing, the dimension d of the volume V of a hyperdimensional cube is $V = L^d$, where L is the length of one of the sides of the hypercube. Consider now the fluctuations δX of some physical quantity X of a turbulent plasma. We may construct some convenient scale-dependent measure μ for the fluctuations – equivalent to the volume of the hypercube – and study how the measure varies with the scale size δ – equivalent to the inverse of the length of one of the sides of the hypercube. If the variation between μ and δ follows that of a power law $\mu \sim \delta^{-d_f}$, then the value d_f may be visualized as the dimension of the fluctuations for the particular chosen measure μ. (Because the usual property of μ increases as δ decreases near a power law singularity, the minus sign is introduced in the definition of d_f.) If further the obtained dimension of the measure of the fluctuations may be understood from ordinary dimensional analysis arguments, one visualizes that the fluctuations can fully occupy the space in which they reside. However, if the obtained dimension of the fluctuations differs from that surmised from dimensional analysis; the fluctuations cannot fully occupy the space in which they reside. Such a dimension has been termed a "fractal" or fractal dimension by Mandelbrot (1977).

It is apparent that the power law relations described in Eqs. (1.4), (1.5), and (1.10) may be interpreted in terms of the concept of fractals because the scaling exponents in these expressions generally cannot be deduced from dimensional arguments.

For intermittent fluctuations, one visualizes that the fluctuations are composed of many types, each type being characterized by a particular fractal dimension. The questions to ask then are (1) what are the different types of fractal dimensions? and (2) how are they distributed in the turbulent medium? These will be one of the central themes of discussion in this monograph, particularly in terms of a rank-ordered multifractal analysis (ROMA).

1.5 Dynamic Renormalization Group

For nonlinear stochastic systems exhibiting complexity, the correlations among the fluctuations of the random dynamical fields are generally extremely long-ranged and there exist many correlation scales (as exhibited, e.g., by the various outsized plasma coherent structures). The dynamics of such systems are notoriously difficult to handle either analytically or numerically. By contrast, because the correlations are extremely long-ranged, it is reasonable to expect the system to exhibit some sort of invariance under coarse-graining scale transformations. A powerful technique that utilizes this invariance property is the method of the dynamic renormalization group (actually a semigroup) as briefly described in the following text.

No matter how complex the stochastic behavior is for the turbulent system, we may conjecture that the system is somehow characterized by a set of dynamical parameters, $\{P_n\}$. We now transform the stochastic system by coarse graining and ask how these parameters evolve accordingly (subject to the rules stipulated by the underlying dynamical equations). Symbolically, we may write

$$\partial\{P_n\}\partial\ell = R\{P_n\} \tag{1.17}$$

where R represents the coarse-graining or renormalization transformation operator and ℓ *is* the coarse-graining parameter. The operator is generally very complicated for even relatively simple stochastic systems.

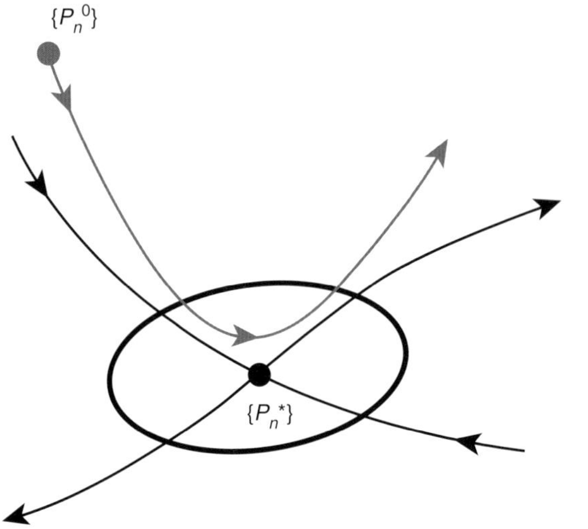

Figure 1.3. Parameter space of a dynamic renormalization group (RG). A typical (gray) RG trajectory initiates from some point $\{P_n^0\}$, skirts close to and remains under the influence of the (elliptical) domain of attraction of the partially repulsive (saddle) fixed point $\{P_n{}^*\}$ before migrating away. Also shown are (black) trajectories (separatrices) that emerge or terminate at the fixed point. (For a color version of this figure, please see the color-plate section.)

In analogy to nonlinear dynamics, we expect that the transformation (1.17) will encompass fixed points (singular points) in the phase space of the parameters, at which $d\{P_n(\ell)\}/d\ell = 0$. For most stochastic systems, fixed points are generally partially repulsive. This is particularly true for nonconservative systems. Example of a partially repulsive fixed point in a two-dimensional parameter space would have the characteristics of a saddle point as graphically depicted in Figure 1.3. At a fixed point, for example, $\{P_n^*\}$, all length scales should not be changing. However, the coarse-graining transformation requires that all length scales change with ℓ. Therefore, to satisfy both requirements, the characteristic correlation length must be either equal to infinity or zero at the fixed point. When the correlation length is infinite, the stochastic system is at criticality.

A renormalized stochastic state upon continued steps of coarse graining may sometimes migrate into a region in the parametric space close to a partially repulsive fixed point (e.g., inside the elliptical region of Figure 1.3) and remain there for many additional iterations before eventually meandering away. Then, in coarse-grained sense, the stochastic system would be close to a state of dynamic criticality. When this occurs, we may approximate the coarse-graining transformation by linearizing the operator R. The mathematical consequence of the linear approximation is that, close to dynamic criticality, certain linear combinations of the deviations of the parameters from the fixed point that characterize the stochastic phenomenon will correlate with each other in the form of power laws – that is, exhibit fractal behavior.

Scaling arising from partially repulsive fixed points involves at least one relevant (repulsive) scaling direction in the parameter space in the sense that the system would be at criticality (i.e., of infinite correlation length) when the parameter characterizing that direction vanishes. And the correlation length becomes finite, decreases in magnitude, and scales with the increase of that parameter as a power law. Such phenomenon as characterized by a partially repulsive fixed point occurs only if there is some tuning of the system parameters such that the renormalized trajectory will evolve close to the fixed point as illustrated by the typical RG path in Figure 1.3.

As the dynamical state deviates away from the vicinity of a partially repulsive fixed point, its fractal behavior changes (*crosses over*) to other fractal states. The parameters that characterize such changes or *crossovers* of fractal behavior generate a spectrum of the multifractals as discussed in Section 1.4.

Sometimes, the coarse-grained trajectories of stochastic systems may evolve into the domain of an attractive fixed point (i.e., a sink singularity). Then, the renormalized stochastic systems characterized by these trajectories all approach a unique stationary state without the tuning of the parameters. And the scaling results (in particular, power-law distributions) are fully characterized by the properties of the attractive fixed point and thus universal. Such then would be the general scenario for the phenomenon of SOC as originally proposed by Bak and colleagues (1987). In practice, SOC states may exist only if the dynamical system is conservative.

1.6 Application to Space Plasmas

Applications of the ideas of dynamical complexity to space plasmas were relatively recent. The first contact of these ideas with space plasmas was probably contained in a series of papers addressing the apparent low-dimensional chaotic behavior related to the dynamics of the Earth's magnetosphere (Baker et al. 1990; Klimas et al. 1992; Roberts et al. 1991; Shan et al. 1991a, 1991b; Sharma et al. 1993). These ideas were followed by a paper by Chang (1992) that suggested the possibility of interpreting such phenomenon from the point of view of forced and/or self-organized criticality (FSOC) surmised from the method of the dynamic renormalization group (DRG). This suggestion was motivated by the low-dimensional chaos papers as well as the interpretation of the observed large variability of plasma fluctuations in the Earth's magnetotail as the result of sporadic and localized current disruptions (Lui et al. 1988). At approximately the same time, Lu and Hamilton (1991) used the idea of SOC of Bak and colleagues (1987) to explain the curious scale-independent behavior of the solar flare occurrence rate on the flare sizes. Such ideas suggest that complex dynamical systems generally organize themselves into states with statistical properties describable by power laws. (Succinct tutorial reviews of the SOC concept may be found in Jensen [1998] and Aschwanden [2013]). An alternative viewpoint of dynamical complexity is based on the concept of multifractals (Halsey et al. 1986; Parisi and Frisch 1985) which is a generalization of the original ideas of fractals introduced by Mandelbrot (1977). Burlaga (1991a; 1991b; 1993) was the first to incorporate such ideas to interpret the intermittent turbulent behavior of the solar wind. Other solar wind studies based on such a point of view followed, (e.g., Carbone 1993; 1994; Carbone et al. 1995; 1996; Carbone and Bruno 1996; Horbury et al. 1995a; 1995b; 1997; Marsch et al. 1996; Marsch and Liu 1993; Rusmaikin et al. 1993). These ideas were followed by Consolini and colleagues (1996) and Consolini (1997) in their interpretations of the observed fractal time series of the Auroral Electrojet (AE) indices.

Since then, a flurry of activities has blossomed into the current investigations of the phenomenon of complexity in space plasmas. Perhaps the most noteworthy developments are the connection of such phenomena with the traditional ideas of intermittent turbulence in plasmas and the development of the concept of crossovers (transitions) from one criticality to another and onto multifractals (Chang and Wu 2007; 2008). This monograph will provide simple descriptions of some of these modern ideas in conjunction with selected actual observational results in space and numerical simulations.

1.7 Outline of the Monograph

In Chapter 2, we consider the phenomenon of dynamical criticality far from equilibrium. Concepts such as FSOC and SOC will be introduced. Observational and numerical examples in solar flares, geomagnetic storms, and substorms exhibiting dynamic criticality in space plasmas will be discussed. Mean field theory, and running models of cellular automata as well as coupled map lattices and direct numerical simulations describing such processes will be introduced. Both spatial and spatiotemporal dynamic structures will be analyzed.

This is followed by a chapter discussing the origin of the physics of complexity in space plasmas. The concepts of coherent structures and their interactions will be introduced. In particular, the basic idea of coarse-grained dissipation will be explained. Chapter 3 ends with a sufficiently detailed description of the method of wavelet transforms geared toward the applications to space plasmas.

Chapter 4 introduces the concept of PDFs of fluctuations involving incremental changes in space and time. Empirical models such as those of Castaing and Kappa will be discussed. The connection between the Kappa distribution and the nonextensive entropy concept of Tsallis will be described. The Taylor hypotheses commonly employed in analyzing space data will be examined and the method of one-parameter rescaling (which is an extension of the FSS) will be explained. The method of structure functions capable of studying intermittency and multifractals then follows. The formalism of Parisi and Frisch incorporating the idea of global and local invariants are introduced. And the empirical method of the log-Poisson analysis of the structure function exponents will be explained.

Chapter 5 starts with the introduction of the concept of partition functions in intermittency analysis. The multifractal definition of the partition function exponent will be discussed. This will be followed by the interesting concepts of generalized dimensions and singular measures and related spectrum. An example related to the intermittent turbulence in the cusp region of the magnetosphere will be considered and analyzed using the wavelet transform modulus maxima (WTMM) method.

Chapter 6 is devoted to the concept, motivation, and development of the method of ROMA. Observational and numerical applications will be provided to explain this new method of intermittency analysis. The connection between the idea of ROMA and local and global invariants as well as the related crossover phenomena will be carefully explained.

Finally, in Chapter 7, the ideas of the DRG will be discussed. Using toy models and simple examples, the usefulness and importance of DRG will be explained. In particular, how such theoretical ideas may be weaved together with the basic concepts of scaling near criticality, FSS, ROMA, local invariants, and crossovers will be described.

A brief "finale" concludes this introductory monograph.

2

Dynamical Criticality Far from Equilibrium and Associated Phenomena

2.1 Introduction to Self-Organized Criticality

In their original paper on self-organized criticality (SOC), Bak and colleagues (1987) suggested that for a dynamical system with many nonlinearly interacting parts its statistically stationary configuration could be organized into a complex state exhibiting properties akin to those for equilibrium critical states with scale-invariant, power-law singularities. And the claim was that such organization could develop without the external tuning of the system parameters. They suggested that the basic requirement for the development of a self-organized critical state was the existence of many metastable states that could be perturbed into unstable states by threshold perturbations and that such local instabilities could propagate throughout the dynamical medium in terms of avalanches of various sizes. A sandpile model of cellular automaton was suggested to demonstrate their basic idea. In this lattice model, if the sandpile was perturbed locally such that the slope at a lattice site became larger than a critical value, the sandpile would collapse there. That local collapse might then cause the slopes of its adjacent lattice sites to exceed the critical value and hence induce new localized collapses. This process continued until the sandpile eventually became everywhere stable. Such an event by definition was called an "avalanche." New avalanches were then created successively by other random threshold perturbations. After many avalanching processes the sandpile system would then reach a state where the statistical distributions of the properties of the avalanches became stationary – an SOC state – and the distributions would be generally characterized by scale-invariant, power laws. Numerical results for the size and duration distributions seemed to verify such an assertion.

Perhaps the first contact with the concept of SOC in space plasma applications was the recognition of the power-law, scale-independent behavior of the occurrence

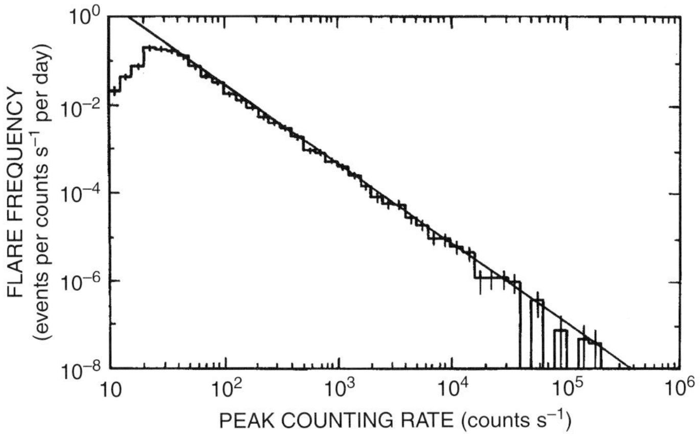

Figure 2.1. Distribution of flare occurrence rate versus peak count rate of all flares detected with HXRBS (Hard X-Ray Burst Spectrometer) from launch to February 1985. The straight line through the data corresponds to a power law with a spectral index of 1.8. (From Dennis 1985.)

rate of solar flares (sudden releases of large amounts of energy with intense radiation and particle acceleration in localized regions of the solar atmosphere) on their sizes, Figure 2.1 (Dennis 1985; Lu and Hamilton 1991). In analogy to the sandpile cellular automaton, Lu and Hamilton constructed a stochastic flare model based on the avalanches induced by many small reconnection (magnetic field-line merging) events. In their model, a three-component vector magnetic field $\mathbf{B}_i$ on a three-dimensional cubic lattice representing the average magnetic field in a cell was specified, where the index i represented the spatial location of the cell. The local magnetic gradient $d\mathbf{B}_i$ was then defined as the difference between the local magnetic field and the average of its six nearest neighbors $\mathbf{B}_{nn}$: $d\mathbf{B}_i = \mathbf{B}_i - (1/6)\sum_{nn} \mathbf{B}_{nn}$. When the magnitude of the local gradient $|d\mathbf{B}_i|$ was larger than a critical value B_c, the configuration was assumed to become unstable locally (due to reconnection or other mechanisms). As this occurred, the local magnetic field would then transfer (1/7) of the local gradient to each of its six nearest neighboring sites reducing the local gradient to zero. The fields at the nearby positions might then become unstable according to the same criterion, resulting in a new distribution of the fields. During each such an instability event, a magnetic energy $(6/7)\,|d\mathbf{B}_i|^2$ was released.

Specifically, in their simulation, the system started with a uniform magnetic field driven at some random position by the addition of a random vector $d\mathbf{B}_i$. If the local magnetic gradient exceeded the critical value, the field was readjusted as discussed in the preceding text. This process continued until the field was everywhere less than the critical value. After such an avalanche, another random vector was added somewhere and this process was repeated. After many such avalanches, the resulting simulated distributions of events became stationary and

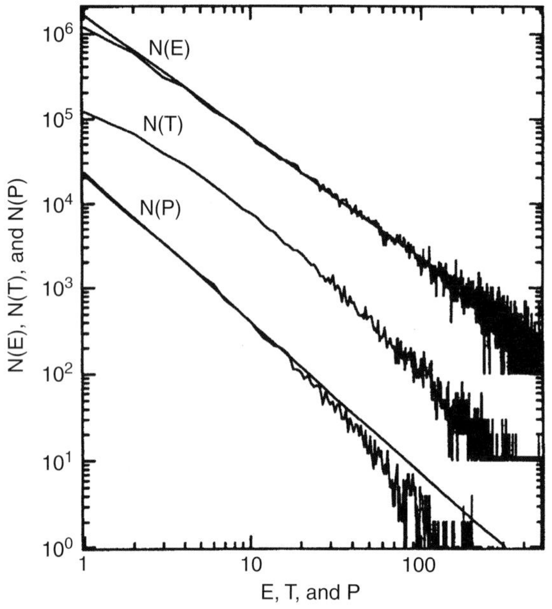

Figure 2.2. Distribution of event energy release $N(E)$, peak flux $N(P)$, and duration $N(T)$ in arbitrary units for avalanches of the Lu-Hamilton model of a 30^3 grid. The power-law relations of $N(E) \sim E^{-1.4}$ and $N(P) \sim P^{-1.8}$ are also shown for comparison. Note that $N(E)$ and $N(T)$ are offset by factors of 100 and 10, respectively. (From Lu and Hamilton 1991.)

the distributions of the total energy, peak flux, and duration (Figure 2.2) exhibited the expected power-law behavior of SOC. In particular, the peak flux distribution exhibited a logarithmic slope of 1.8, which was in very good agreement with the observed result of Figure 2.1. And, the results were insensitive to the choice of B_c, nor the redistribution law, provided the sum of the magnetic field vectors was conserved.

These ideas on solar flares were refined and generalized by Vlahoes and colleagues (1995) and Einaudi and Velli (1999). After this initial activity, there has been a flurry of studies of SOC and other related critical phenomena for dynamical space plasma systems far from equilibrium. (See, e.g., the space plasma review articles in *Self-Organized Criticality Systems*, edited by Aschwanden 2013.) We shall address a few of such applications in observations, theory, and numerical simulations in Section 2.4.

2.2 Flicker Noise and Running Avalanche Models

The $1/f^\beta$ ***and*** $1/f^2$ ***Spectra.*** It has long been noticed that the power spectra of the time series of many naturally occurring phenomena exhibit the "$1/f$" or more generally "$1/f^\beta$" behavior in the low-frequency limit with a nontrivial β, where f denotes frequency (see, e.g., Press 1978). Bak and colleagues had suggested that the power spectrum (square of the amplitude of the Fourier transform) of the total instantaneous dissipation rate for their original model could explain this "flicker noise" phenomenon. However, it was soon realized (Jensen et al. 1989; Kertész and Kiss 1990) that such an expectation had its limitations. In fact, it may be easily demonstrated that for noninteracting avalanches of the Bak and colleagues (1987, 1988) model, the power spectrum would generally have a low frequency behavior of "$1/f^2$," not "$1/f^\beta$."

Running Sandpiles

Noting the preceding deficiency, Hwa and Kardar (1992) proposed a sandpile model with transport by depositing sand grains in the cellular automaton system at some constant finite rate (with that of the original Bak et al. model being the limit of zero deposition rate). For a finite nonzero constant rate of deposition and after the system reached a steady state at which the input and the output of sands drained from the open end balanced, the power spectrum of the output current (the rate of the number of sand grains leaving the system) was computed. It was noted that at sufficiently high frequencies, the power-law scaling behavior of the distributions of monitored quantities had $1/f^2$ characteristics. This was attributed to the fact that at this high-frequency regime, the observation time was of the order of the average avalanche duration. Thus most of the avalanches did not yet have the opportunity to interact. Consequently, this portion of the power spectrum exhibited the "$1/f^2$" behavior.

For frequencies below this high-frequency range, the avalanches started to interact. And true dynamical complexity began to set in with the power spectrum starting to exhibit the "$1/f^\beta$" or flicker noise signature. Below frequencies of this intermediate range, the calculations exhibited avalanche sizes of the order of the system size. Within this discharge-event range, "finite size scaling and other multifractal phenomena" (Kadanoff et al. 1989) are generally observed, which we shall discuss in detail in the later chapters of this treatise.

2.3 An Interlude – Mean Field Theory and Scaling Near Criticality

2.3.1 Mean Field Theory

Before we launch into the discussion of specific sample models of criticality with space plasma applications, let us first introduce some basic concepts of dynamic criticality in terms of a mean field theory (i.e., some sort of an approximation theory based on the properties of and interactions with the "averages" of the field variables). Depending on the circumstances, the results characterized by a mean field theory may or may not always be quantitatively accurate. Such a theory, nevertheless, can usually provide some useful intuitive feelings of the basic physics underlying the more accurate properties of the discrete or continuum models such as those to be considered in the following subsections. Our discussions of the mean field theory shall follow closely with the ideas of the single-site approximation advocated by Vespignani and Zapperi (1997, 1998).

Let us consider a dynamical lattice model. For each site "i" on the lattice, we may characterize its dynamical state at each time step by an occupation variable σ_i which can assume q values (e.g., 1 = active, 2 = stable, and 3 = critical for a typical sandpile model). The probability distribution $P(\sigma,t)$ of the complete set of $\sigma = \{\sigma_i\}$ on the lattice and time t (taken as a continuum) generally satisfies a master equation:

$$\partial P(\sigma,t)/\partial t = \mathrm{N}\{P(\sigma,t),\sigma,t\} \tag{2.1}$$

where N is a nonlinear operator that determines the underlying stochastic dynamics of the model. A standard method of attacking such a nonlinear operator equation is through the expansion of $P(\sigma,t)$ in terms of the cluster probabilities $P_n(\sigma,t)$ of n-sites. For a homogeneous system, the cluster probabilities are independent of the positions, with P_1 characterizing the average density of sites in a certain state and $P_{n>1}$ the correlation properties of the stochastic system. The time evolution equations for the cluster probabilities form a hierarchical set that generally is not closed. One procedure to close such set of equations is by truncation: that is, the n-site approximation. In the succeeding discussion, we follow closely Vespignani and Zapperi (1997, 1998) and adopt the lowest order one-site or density-field approximation.

Such a single-site theory highlights the importance of the conversion of the dynamical state (from one type to another) of each site. The theory relegates the detailed nonlocal interactions that cause the changes of the dynamical states to a secondary role. Thus, when averaged over the lattice, the one-site mean field theory is primarily a nonlinear chemical reaction-like theory composed of a coupled set of nonlinear evolution equations describing the changes of the average densities of the dynamical states. This prescription preserves the essence of the underlying universality concept of the SOC approach as advocated by Bak and colleagues (1978). Because the mean field theory neglects the effects of fluctuations, generally it provides only a qualitative picture of the basic stochastic phenomenon.

Denoting the average densities over the sites as ρ_κ ($\kappa = 1,\ldots,q$), we have a set of evolution equations for the one-site approximation

$$\partial\rho_\kappa/\partial t = F_\kappa(\rho_1,\ldots,\rho_q) \tag{2.2}$$

where $F_\kappa(\rho_1,\ldots,\rho_q)$ is a nonlinear function of ρ_κ. We shall choose the normalization such that $\sum \rho_\kappa = 1$. Thus, Eq. (2.2) generally contains $(q-1)$ independent coupled equations.

To allow the following discussions to be more specific, let us refer to a sandpile model as an example with: $q = 3$ (κ = 1, active; 2, stable; 3, critical). Because the densities are smaller or equal to one, we may formally express the functions F_κ in polynomial expansions of the densities. In order to allow for stationary states to exist, we shall set the constant terms in the expansions of F_κ equal to zero. We can conceptually write down the first order terms of F_κ by simple reasoning. For example, for F_1, the coefficient for the first power of ρ_1 is -1 because the active sites become stable at the unitary rate; the coefficient for the first power of ρ_2 is 0 because the stable sites never become active; and the coefficient for the first power of ρ_3 is the energy input per site h (with hL^d being the total energy input per time step, L characterizing the lattice size, and d the dimension of the lattice) because critical sites become active due to the energy input. For $h \to 0$, we have the one-site mean field theory of the original BTW (Bak, Tang, and Wiesenfeld) model, and for nonzero h, we have that of the running sandpile model of Hwa and Kardar.

There is only one bilinear interaction term in F_1. This term arises due to the creation of an active site from a critical site because of the relaxation of its nearest neighboring active site(s). We may write this term as $(g-\varepsilon)\rho_3\rho_1$, where g is the interaction rate that depends on the energy and geometry involved during the relaxation process and ε is the energy dissipated at each site during the relaxation process. Combining these results together, we have

$$F_1 = -\rho_1 + h\rho_3 + (g-\varepsilon)\rho_3\rho_1 + O(h\rho_1, \rho_1^2) \tag{2.3}$$

We note that the average energy dissipated per site ε has two contributing parts: one is the actual dissipation during the relaxation process and the other is an effective dissipation due to the boundary conditions. We shall make these effects more specific when we consider actual sample models.

Similarly, we may conceptually write down the leading terms of F_2 as follows.

$$F_2 = \rho_1 - uh\rho_2 - u(g-\varepsilon)\rho_2\rho_1 + O(h\rho_1, \rho_1^2) \tag{2.4}$$

For obvious reasons, the leading terms have opposite signs of those for (2.3). In addition, we have multiplied the second and third terms by a scale factor u because only a fraction of the stable sites receiving energy will contribute to the stable $\to$ critical process. Thus, to the lowest relevant orders, the dynamical equations for the three densities are:

$$\begin{aligned} &\partial\rho_1 / \partial t = -\rho_1 + h\rho_3 + (g-\varepsilon)\rho_3\rho_1 \\ &\partial\rho_2 / \partial t = \rho_1 - uh\rho_2 - u(g-\varepsilon)\rho_2\rho_1 \\ &\rho_1 + \rho_2 + \rho_3 = 1 \end{aligned} \tag{2.5}$$

For stationary dynamical behavior of such sandpile models, we expect the input and output of energies to balance (i.e., $h = \varepsilon\rho_1$) and we may show that to the lowest order in h, $u = 1/(g-1)$, and therefore

$$\begin{aligned} &\rho_1 = h/\varepsilon \\ &\rho_2 = (g-1)/g + O(h) \\ &\rho_3 = 1/g + O(h) \end{aligned} \tag{2.6}$$

For the original sandpile model of Bak, Tang, and Wiesenfeld (1987), the driving (i.e., the addition of grains of sand) is carried out after each avalanche process has completed. Thus, the process of driving is infinitely slow, that is, it is carried out in the limit of the ordering field $h \to 0$. In addition, there is no internal dissipation during the avalanching processes. The only dissipation is that due to the loss of sand grains through the boundaries. In this case, the control parameter (relevant scaling field) is the boundary dissipation, ε. The order parameter of the process is easily recognized from (2.6) as ρ_1.

The preceding discussion is couched on the idea of a sandpile model. It is obvious that similar approaches may be employed with respect to other stochastic physical systems or models. We shall have occasion to consider, for example, the mean field ideas for the modified forest fire model, magnetic fluctuations of the field reversal model, statistical distributions of global auroral images, and so forth in the ensuing sections and chapters.

2.3.2 Scaling Theory

Once the order parameter, the ordering field, and control parameter are identified for a stochastic system exhibiting dynamic criticality, we may proceed to consider its scaling behavior in analogy to the ideas discussed in Chapter 1 related to equilibrium thermodynamic systems.

For example, we may define the exponent δ to characterize the scaling behavior between the order parameter and the ordering field, and the exponent γ to characterize the scaling behavior of the zero ordering field susceptibility (i.e., the response of the change of the ordering parameter due to the imposition of the ordering field) with respect to the control parameter, in analogy to the notations employed for equilibrium critical phenomenon used in Eqs. (1.4) and (1.5).

Let us demonstrate this using the mean field results of Eqs. (2.6) of the sandpile model. From (2.6), we note that because $\rho_1 \leq 1$, the right-hand side of the first expression, h / ε must be ≤ 1. Therefore, the critical state of $\rho_1 = 0$ must be approached in the double limit of $h, \varepsilon \to 0$ and $h / \varepsilon \to 0$. For $h > 0$ and $\varepsilon > h$, the order parameter ρ_1 scales with h linearly, and we have the supercriticality behavior

$$\rho_1 \sim h^{1/\delta} \quad \text{with} \quad \delta = 1 \tag{2.7}$$

And for $h \to 0$ and $\varepsilon > 0$, we have the subcriticality behavior of

$$\chi = \varepsilon^{-\gamma} \quad \text{with} \quad \gamma = 1 \tag{2.8}$$

In addition, there are other scaling exponents that we may define for a system exhibiting dynamic criticality. For example, it may be useful to study the probability distributions of the sizes s, energy densities E, lifetimes T, and decay of the avalanches (if they can be defined) as follows:

$$P(s) \sim s^{-\tau_s}, \quad P(E) \sim E^{-\tau_E}, \quad P(T) \sim T^{-\tau_T}, \quad \rho_1 \sim t^{-\tau_p} \tag{2.9}$$

where the τ's are the corresponding scaling exponents.

Cutoff Functions for Dynamic Systems Near Criticality

For a system at stationary dynamic criticality, we may define a d-dimensional radial response function $\chi(r) \sim r^{d-2+\tilde{\eta}}$ (for $h \to 0$) such that the total current $j = \partial\chi(r) / \partial r$ across a $(d-1)$ hypersurface is conserved, where $\tilde{\eta}$ is the anomalous dimension of the response function analogous to the correlation function exponent in thermodynamic critical phenomenon. The anomalous dimension $\tilde{\eta}$ vanishes for the mean field theory and $\chi(r) \sim r^{2-d}$.

In the presence of dissipation, the dynamic system is not exactly at criticality and acquires a finite correlation length. It is then conventional to express the zero field response function in terms of a cutoff function as:

$$\chi(r) = r^{2-d+\tilde{\eta}} C_\chi(r / \xi) \tag{2.10}$$

where ξ is the correlation length and $C_\chi(r / \xi)$ has the cutoff effect of decaying rapidly for $r > \xi$ and remaining constant otherwise.

For mean field theory $\tilde{\eta} = 0$. Integrating (2.10) over $r^{d-1}dr$, we obtain the susceptibility $\chi \sim \xi^2$. Assuming $\xi \sim \varepsilon^{-\nu}$ where ε is the control parameter (or relevant scaling field) and ν the correlation length exponent, we obtain from the scaling law (2.8) for the sandpile model, $\nu = \gamma / 2 = 1/2$.

We may analogously consider other scaling relations and scaling exponents near criticality following the preceding arguments in similar manners. For example, we may define the probability distribution functions of (2.9) as:

$$P_s(s,\varepsilon) = s^{-\tau_s} C_s(s / s_c(\varepsilon)) \tag{2.11a}$$

$$P_s(E,\varepsilon) = E^{-\tau_E} C_E(E / E_c(\varepsilon)) \tag{2.11b}$$

$$P_T(T,\varepsilon) = T^{-\tau_T} C_T(T / T_c(\varepsilon)) \tag{2.11c}$$

$$\rho_1 = t^{-\tau_p} C_p(t / t_c(\varepsilon)) \tag{2.11d}$$

with $s_c \sim \varepsilon^{-\nu_s}$, $E_c \sim \varepsilon^{-\nu_E}$, $T_c \sim \varepsilon^{-\nu_T}$, and $t_c \sim \varepsilon^{-\nu_t}$ where C_s, C_E, C_T, and C_t are the cutoff functions, ν_s, ν_E, ν_T, $\nu_t = z\nu$ the corresponding characteristic exponents and z the dynamic exponent.

All these introduced exponents are not entirely independent of each other, we may obtain the exponent equalities relating them by following the type of invariant and generalized homogeneous function arguments that we employed for the critical exponents of equilibrium thermodynamics as described in Chapter 1.

In addition, for finite size systems near criticality, the correlation length is comparable to the linear dimension L of the system size. Thus expressions such as those of (2.11) may be conveniently translated into finite-size scaling (FSS) forms by relating ε to ξ and then in terms of L.

The preceding discussions may easily be extended to other critical dynamic systems by replacing the dissipative (control) parameter with the general relevant scaling field in the expressions. We shall have the opportunity to apply such transformed FSS relations in Section 2.4.3.

We conclude this subsection by noting that the scaling laws and the related exponent equalities are quite general and are, in fact, independent of the mean field arguments.

2.4 Scaling Properties of Global Geomagnetic Fluctuations

2.4.1 Introduction

The Sun emits a "solar wind" of charged particles originating outward from the solar coronal holes and traveling along the open magnetic field lines with flow speeds at hundreds of kilometers per second. As the solar wind impinges upon the Earth, the Earth's magnetic field carves out a semi permeable bubble like structure from it – the Earth's magnetosphere – extending several Earth's radii on the dayside and many radii on the nightside. The behavior of the magnetosphere is extremely complex. Two of the most discernible phenomena associated with the global dynamics of the magnetosphere are the "substorms" and "storms."

A *geomagnetic substorm* is a brief disturbance in the Earth's magnetosphere that causes energy to be released from the "tail" of the magnetosphere and injected into the high latitude ionosphere. Discrete aurorae (light displays at the high latitude regions caused by energetic particles colliding with atoms in the atmosphere) are commonly generated during magnetic substorms. The substorm may generally be divided into three dynamical phases: the growth phase, the expansion phase, and the recovery phase.

(1) The growth phase is commonly assumed to be initiated by the southward (i.e., in the opposite polarity of the dayside geomagnetic field) turning of the interplanetary magnetic field permitting the field lines to merge at the dayside magnetopause. This in turn allows energy to be transferred from the solar wind to the magnetosphere and facilitates the storage of some of this energy into the nightside region (the magnetotail). The magnetic energy content of the magnetotail thus increases during the growth phase and this process somehow causes the thinning of the denser central plasma region of the magnetotail (the so-called plasma sheet).

(2) The expansion phase onset represents a switch from relatively slow rate of storage of energy in the magnetotail of the growth phase to a very rapid rate

of dissipation and release. This process is generally assumed to be initiated by the reconnection (magnetic field line merging) in the near Earth magnetotail region of the thinned plasma sheet. It is characterized by the injection of energetic particles in the high latitude ionosphere causing visible auroral displays and joule heating by the increased ionospheric currents.

(3) The substorm terminates with the recovery phase, in which the magnetosphere relaxes back into its quiet time configuration.

One of the more popular observational time series indices that has proven to be useful for the monitoring of the intensity of the global geomagnetic activities and, in particular, the substorm activities, is the AE (auroral electrojet) index (Davis and Sigiura 1966). It is derived from the observations of the horizontal component of the Earth's magnetic field, obtained from ground-based stations situated in the high-latitude regime (i.e., the auroral zone). It provides a measure of the overall horizontal current strength of the AEs. These electrojets are Hall currents carried primarily by electrons at altitudes from 100 to 150 kilometers. They are connected to a system of current filaments that flow along geomagnetic field lines into and away from the polar region. Therefore, the AE index provides a useful gauge of the geomagnetic substorm activities.

A *geomagnetic storm* may be caused by the interplanetary disturbance due to the solar event of a coronal mass ejection (CME), a corotating interaction region (CIR) of the fast and slow solar wind streams, or other interplanetary fluctuations. Most of the magnetic storms are composed of three phases: an initial phase, a highly disturbed main phase, and a recovery phase. An index Dst (storm time disturbance) is popularly used to gauge the strength of geomagnetic storms. It is based on the average global horizontal component of the Earth's magnetic field of stations situated at the magnetic equator. Quiet time readings of Dst fluctuate between plus and minus 20 nT (nano-Tesla). The initial phase of the storm generally involves fast increases of Dst from 20 to 50 nT in tens of minutes. The main phase with a duration of several hours is characterized by Dst values from –50 nT to minus hundreds of nT, while the recovery phase may last from hours to several days. The AE and Dst indices are related, but generally not in a straightforward linear fashion.

In a well-referenced paper, Consolini (1997) studied a data set of the AE-index with one-minute time resolution covering the whole period of the year 1978 including various global geomagnetic disturbances. He studied the distribution function for the burst events of the time series by defining the strength s of a burst as:

$$s = \int_{\Omega} (Ae(t) - L_{AE})dt \tag{2.12}$$

where the threshold cutoff L_{AE} is chosen the quiet time level of [45 ± 15] nT. The integration is performed over the time interval Ω where the condition $Ae(t) \geq L_{Ae}$ is satisfied. The definition of the burst strength s may be related to the magnetospheric energy dissipated by the AE.

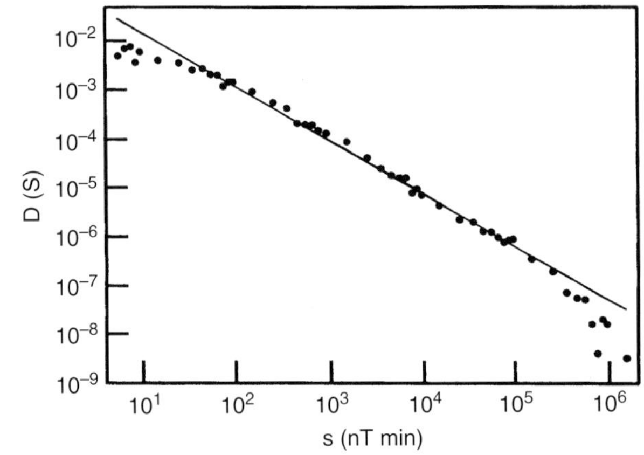

Figure 2.3. Distribution $D(s)$ of the burst size $s = \int_{\Omega} [Ae(t) - L_{AE}]dt$ of an AE index time series, where the integration is performed over the time interval Ω when the quiet level of $L_{AE} = (45 \pm 15)\text{nT}$ is exceeded. The solid line is a power-law best fit. (From Consolini 1997.)

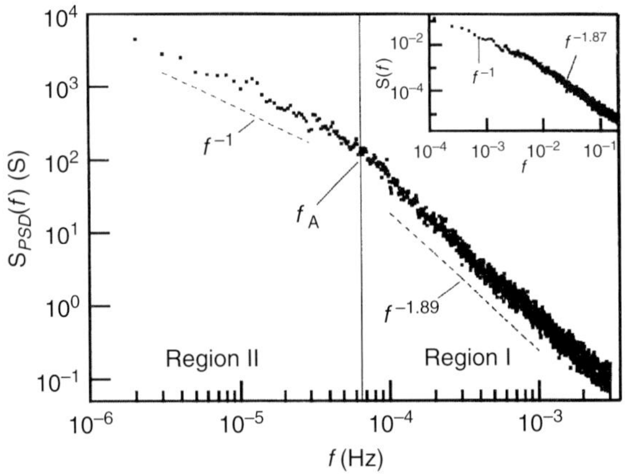

Figure 2.4. Power spectrum density (PSD) of the entire AE series analyzed by Consolini. The log-log plot shows a broken power-law fit with a spectral break at $f_A \sim 6.3 \times 10^{-5} Hz$ and exponents of approximately –1 and –2. The insert is the PSD obtained from a running sandpile cellular automaton. (From Consolini 1997.)

The log-log plot of the burst distribution $D(s)$ for a subset of the considered AE series containing nearly three thousand events indicated a power-law fit with an exponent of –1 over more than four decades, Figure 2.3, displaying a signature of dynamic criticality.

In addition, the power spectrum of the entire AE series is obtained, Figure 2.4. The log-log plot showed a broken power-law fit with a spectral break at $f_A \sim 6.3 \times 10^{-5} Hz$ and exponents of approximately –1 and –2. The lower frequency range with an exponent of –1 represent the typical duration range of magnetic storms and sub-storms (lasting from hours to days) and can be interpreted from the point of view

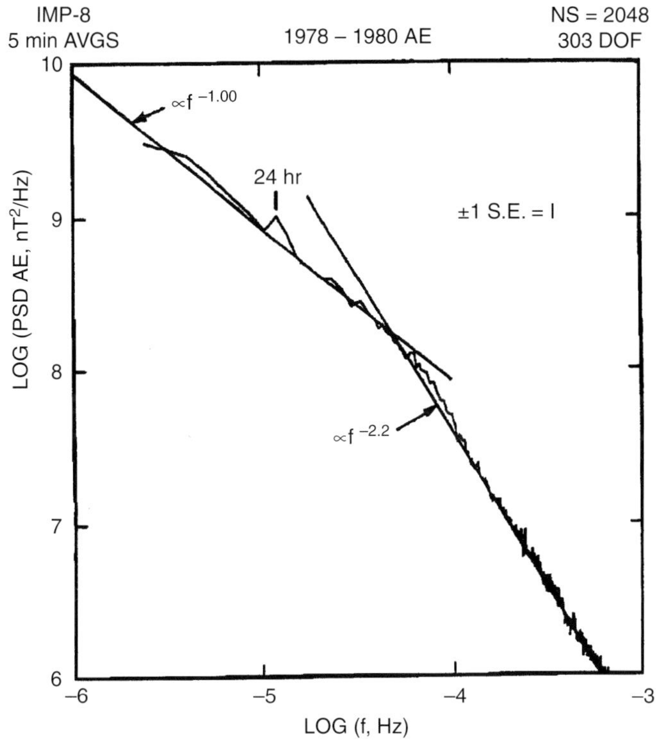

Figure 2.5. Power spectrum of five-minute averages of AE index for years 1978 through 1980. (From Tsurutani et al. 1990.)

of running sandpiles of overlapping avalanches (bursts) as discussed by Hwa and Kardar (1992). The high-frequency range would then represent regimes when the bursts were not overlapping.

In fact, the sharp break in the power spectrum of the AE index was already noted by Tsurutani and colleagues (1990) during their comprehensive study of the nonlinear response of the AE index to the southward component of the solar wind driver, Figures 2.5 and 2.6. The exponents had similar values as those found by Consolini (1997) and the breaks corresponded to a time period of approximately five hours.

2.4.2 Global Auroral Images and Scaling

Lui and colleagues (2000) were the first to consider the statistical properties of global auroral images at all scales from the vantage point of a satellite to study the energy output of the magnetospheric dynamics. They examined 9,033 frames of ultraviolet images collected by the UVI (Ultraviolet Imager) instrument on the Polar spacecraft (spacecraft launched on February 24, 1996 designed to study the polar magnetosphere and aurora) for the period of January 1–31, 1997, Figure 2.7. The period included a major magnetic storm on January 10–11, 1997 and a number

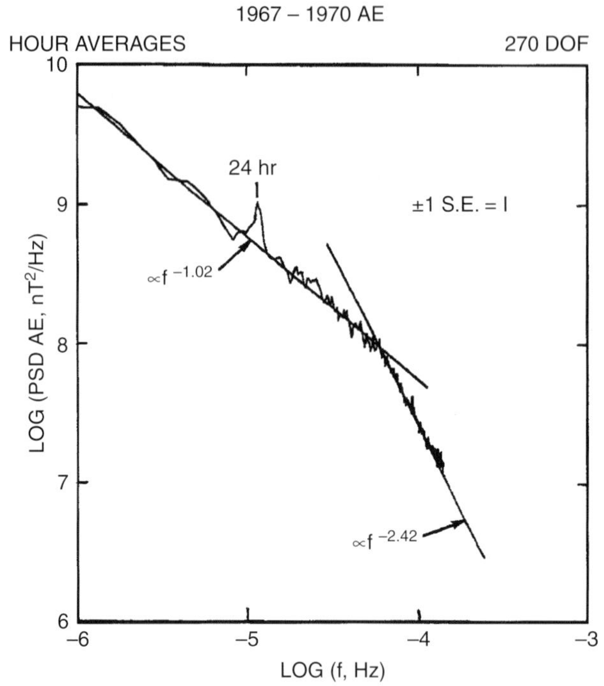

Figure 2.6. Power spectrum of one-hour averages of AE index for years 1967 through 1970. (From Tsurutani et al. 1990.)

of magnetic substorms. Only nighttime observations were used to eliminate the uncertainties of the dayglow.

To determine the spatial distribution sizes (areas) of the auroral activity regions, the background intensity outside of the luminous auroral belt was determined and subtracted from the intensity of the auroral belt. Individual auroral blobs were then identified by applying a threshold intensity level that is well above the instrument noise level. (It was found that the statistical results were not sensitive to the choices of threshold level.) The total number of identified auroral blobs for this study included 81,794 for quiet times and 155,347 for magnetic substorm intervals. After the sizes of the blobs were obtained (taking into account the wobbling effect on the auroral images due to the nutation of the satellite spin axis), the dissipated power of each blob was calculated according to a procedure described by Germany and colleagues (1997). The normalized distributions of size in log-log plots are given in Figure 2.8 a, b. For quiet time intervals, it exhibited a power-law distribution with slope of -1.161 ± 0.033. For substorm intervals, the distribution showed a prominent peak above $\sim 10^5 \mathrm{km}^2$. Between $\sim 5 \times 10^3$ and $\sim 5 \times 10^4 \mathrm{km}^2$ there was an indication of a power law with a slope of -1.206 ± 0.079. Thus, within the bounds of uncertainties the two power-law slopes were similar.

The normalized distributions for power exhibited similar properties. For quiet time, the distribution had a power-law slope of -0.995 ± 0.022, Figure 2.8 c, d. There was a peak for the substorm interval distribution for powers exceeding 5×10^8

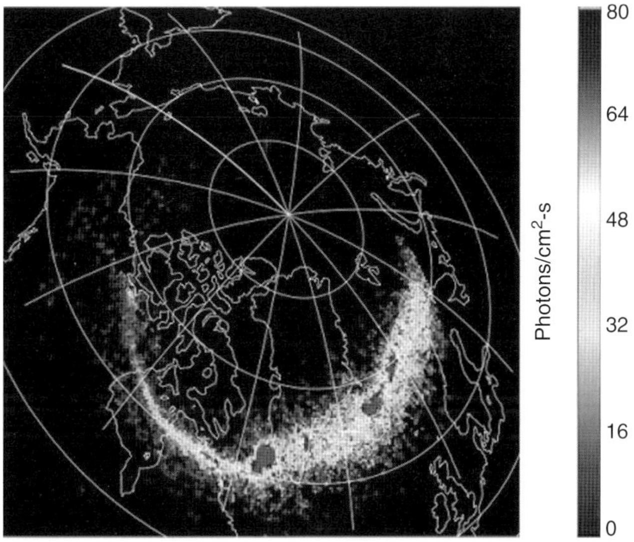

Figure 2.7. A sample of global auroral distribution from the UVI imager of the Polar spacecraft. (From Lui et al. 2000.) (For a color version of this figure, please see the color-plate section.)

Watt. Between $\sim 5\times10^6$ and $\sim 5\times10^8$ Watt, the distribution has a power-law slope of -1.049 ± 0.083. These results remained unchanged with other thresholds used in the analysis.

Lui and colleagues (2000) interpreted their results of similar power-law exponents as an indication of dynamic criticality and the peaks in the distribution as the manifestation of geomagnetic global reconfiguration. Their interpretation was motivated by the comparison of their observational results with an avalanche model based on a one-dimensional sandpile automaton (Chapman et al. 1998; 1999; Watkins et al. 1999).

Their original one-dimensional sandpile model involved a grid of N equally spaced cells of sand. The sandpile was assumed to have an "angle of repose" below which it was always stable. The heights and gradients of sand were measured relative to the angle of repose. Each cell was assigned a critical gradient selected randomly from a "top hat" probability distribution. Sand was added at cell 1 at some rate and is redistributed to cell 2 if the critical gradient was exceeded. The redistribution was conservative and instantaneous. If the local random critical gradient was exceeded again, the redistribution propagated to cell 3 and so on. Within each avalanche, the sand was instantaneously "flattened" back to the angle of repose. The critical gradients at cells within the flattened avalanche region were then regenerated and more sand was added at cell 1 until it was unstable, triggering another avalanche. An avalanche might be entirely an internal arrangement or continue until it spread across all cells (a system wide global discharge). The relaxation rule adopted here allowed propagation of information (correlation) across the avalanche and therefore

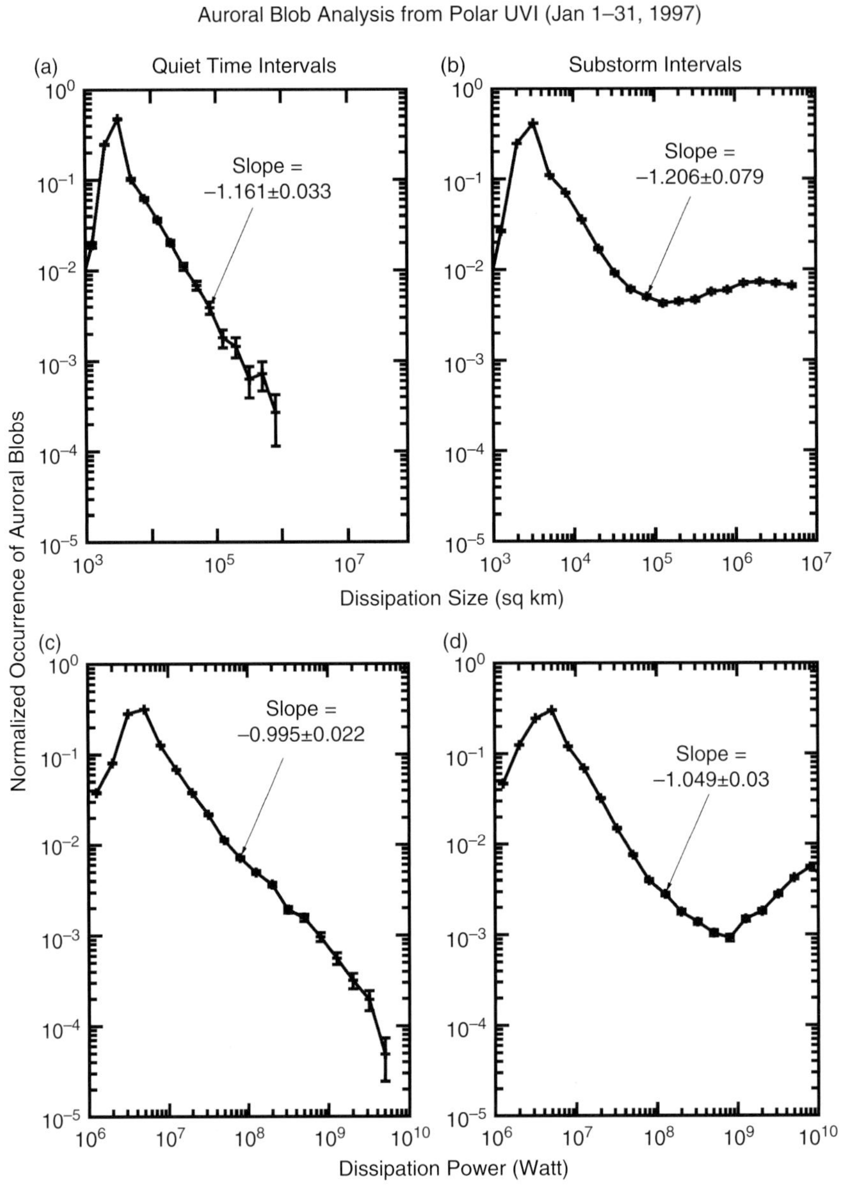

Figure 2.8. Probability distributions of size and power output of an individual auroral region. (a) Size distribution during quiet times. (b) Size distribution during substorms. (c) Power distribution during quiet times. (d) Power distribution during substorms. (From Lui et al. 2000.)

potentially on all length scales in the sandpile allowing dynamic critical behavior in a one-dimensional system, whereas other reorganization rules (e.g., Bak et al. 1987; Jensen 1998) generally have required at least two dimensions.

For a direct comparison, the preceding avalanche model was modified to run with a fluctuating input to mimic the driving of the solar wind. In Figure 2.9, the

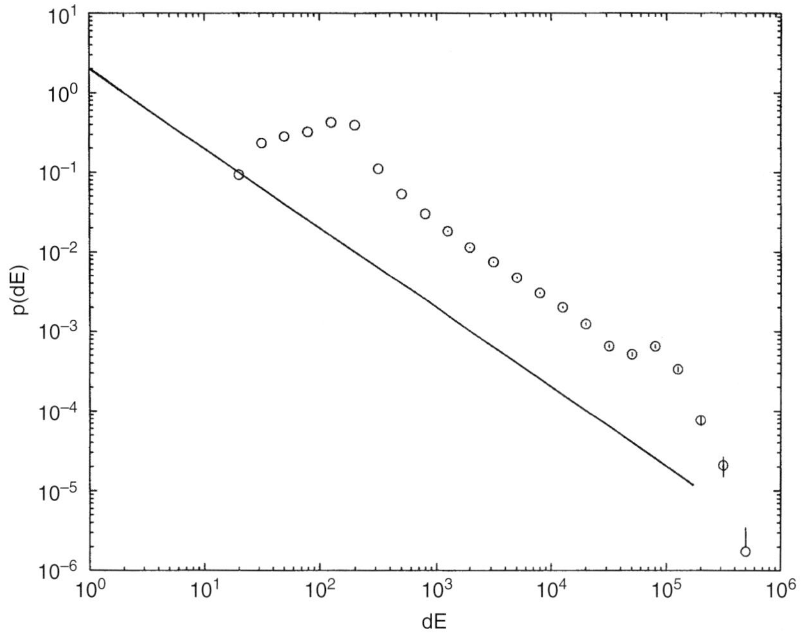

Figure 2.9. Probability distribution of the energy dissipated per avalanche in arbitrary units of a 1D sandpile model of Chapman et al. (1999). The internal avalanches have a power-law distribution with no intrinsic scale whereas the systemwide avalanches show a characteristic scale. The solid line has a slope of −1 for comparison. (From Lui et al. 2000.)

outputs in terms of probability distribution of the total energy dissipated (defined as the difference of the sum of the square of the heights before and after an avalanche) for all events, both internal and global, were shown on a single plot (Lui et al. 2000). The internal events fell on a power law that maintains the same slope independent of the driving rate. The effect of fluctuations in the input was to distort the lowest energy events giving an increase in slope. And the global events appear as a statistically significant peak at the high energy end. Both are similar to those exhibited by the auroral imaging results interpreted as spatial avalanches.

2.4.3 Spatiotemporal Avalanches and Their Scale-Free Statistics

Global Auroral Images

Uritsky et al. (2002a) suggested that the statistics of global auroral images may perhaps be better understood in terms of spatiotemporal avalanches. Their test study was based on an analysis of more than thirty thousand Polar UVI images obtained in January–February 1997 and January–February 1998. The images represent the 165.5 to 174.5 nm portion of the N_2 Lyman-Birge-Hopfield (LBH) spectral band which characterizes the energy flux of auroral electrons precipitating into the high-latitude ionosphere (Doe et al. 1997; Germany et al. 1994). Continuous sequences of images of the nighttime sector were studied. The nutation effect of wobbling of the satellite and thermal noise were taken out and a luminosity threshold was used to insure the detection of the active region of the aurora.

During disturbed times, the active auroral areas can split and/or overlap constituting complex dynamical patterns. These uncertainties were taken care of by the definition of avalanches used in running cellular automaton models of SOC (Becker et al. 1995). Namely, the split perturbations with a unique source were considered parts of a single event, and the merged perturbations with spatially distinct origins were treated as separate ones (the common tail following the merging event was ascribed to the perturbation that started earlier). In total, about 12,300 events were detected.

To determine the spatiotemporal characteristics of the identified events (avalanches), the integrated size S and energy E over the lifetime T were calculated.

$$S = \int_{\{T\}} s(t)dt; \quad E = \int_{\{T\}} w(t)dt \tag{2.13}$$

where $s(t)$ and $w(t)$ were the area of the perturbation and the energy deposition into this area by auroral electrons at time t. Remarkably, normalized occurrences (probability distributions) of T, S, and E for all the events detected during the four studied months had stable power-law forms for different choices of activity thresholds as shown in Figure 2.10 a, b, c. The distributions did not show the "bumps" of the probabilities of spatial auroral events (avalanches) as discussed in the previous section, thus, providing an alternative explanation that the bumps might be due to the overcounting of occurrences of large and prolonged auroral events.

The preceding observed power-law statistics are consistent with the viewpoint that the dynamic magnetosphere operated at stationary criticality.

Finite Size Scaling

Uritsky and colleagues (2006) also demonstrated that the auroral images support the idea of finite size scaling as discussed in Chapter 1 of the equilibrium critical phenomena and Section 2.3 of the scaling theory of critical dynamics. Because of the irregular and anisotropic shapes of the active emission regions, the linear scale l was not always defined. These authors decided to study the finite size effect in terms of the emission area s. Then, assuming that the outer boundaries of the active regions had a fractal geometry, the relationship between l and s might be related as $l = as^{1/d_f}$. They determined that the fractal dimension $d_f = 1.54 \pm 0.02$ and the constant $a \approx 0.281$, by box counting (Turcotte 1997), Figure 2.11.

For the 15,500 UV images obtained during the period of January 1–February 28, 1997, the authors found that the probability distributions $P(E)$ and $P(T)$ of

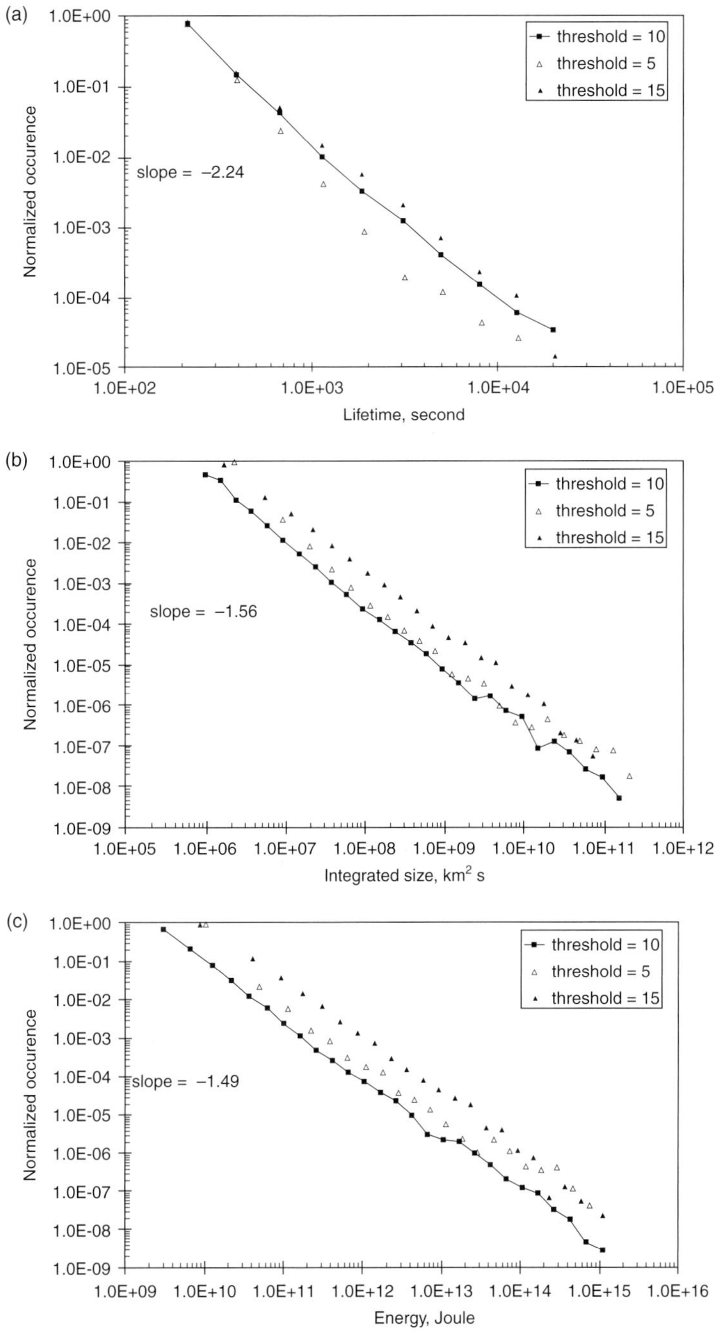

Figure 2.10. Probability distributions of the lifetime, integrated size, and energy per avalanche of the entire spatiotemporal Polar UVI study of the auroral perturbations using three levels of activity thresholds. The slope exponents are estimated from the threshold of 10 $photons \cdot cm^{-2} \cdot s^{-1}$. (From Uritsky et al. 2002a.)

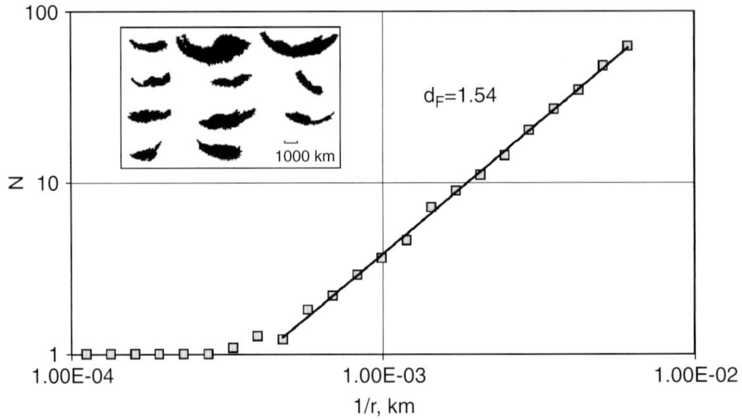

Figure 2.11. Box-counting statistics of twenty representative spatiotemporal Polar UVI auroral events (number of nonoverlapping square boxes required to cover the emission region as a function of the inverse linear box size r). The power-law fit indicates a scale-invariant fractal shape with $d_f = 1.54 \pm 0.02$. Inset: Examples of auroral emission regions included in the statistics. (From Uritsky et al. 2006.)

auroral emission events of energy E and lifetime T from the LBH band may be rescaled as

$$P(E, l < L) = E^{-\tau_E} P_E(E / L^{D_E}) \tag{2.14}$$

$$P(T, l < L) = T^{-\tau_T} P_T(T / L^{D_T}) \tag{2.15}$$

where $\tau_E \approx 1.60$, $\tau_T \approx 2{,}24$, $D_E \approx 2.14$, and $D_T \approx 1.02$.

Figure 2.12 a, b give the rescaled P_E and P_T confirming their approximate data collapsing and cutoff properties, thus, revealing the near scale-free nature of the spatiotemporal auroral emission dynamics.

Dynamic Criticality in the Solar Corona

Uritsky and colleagues (2007) have also studied the spatiotemporal avalanches of solar emissions from SOHO (Solar and Heliospheric Observatory launched on December 2, 1995) in extreme ultraviolet (EUV) range based on the FE XII emission images taken by the EUV imaging telescope (EIT) in the 195 Å wavelength using the same technique as that discussed for the auroral images. The data included two observation periods: June 29, 2001–July 28, 2001 (3,240 images at solar maximum with average sunspot number of 64.0) and October 22, 2005–December 2, 2005 (4,407 images at solar minimum with average sunspot number of 16.3) with a typical time resolution of 13.3 minutes. Contiguous spatial regions $w(\mathbf{r}, t) > w_a$ of the EIT luminosity larger than the activity threshold w_a were treated as evolving avalanches and the spatiotemporal lifetime T, the radiation emission flux $E = \int w(\mathbf{r}, t) d\mathbf{r} dt$, and the peak area A for the avalanches

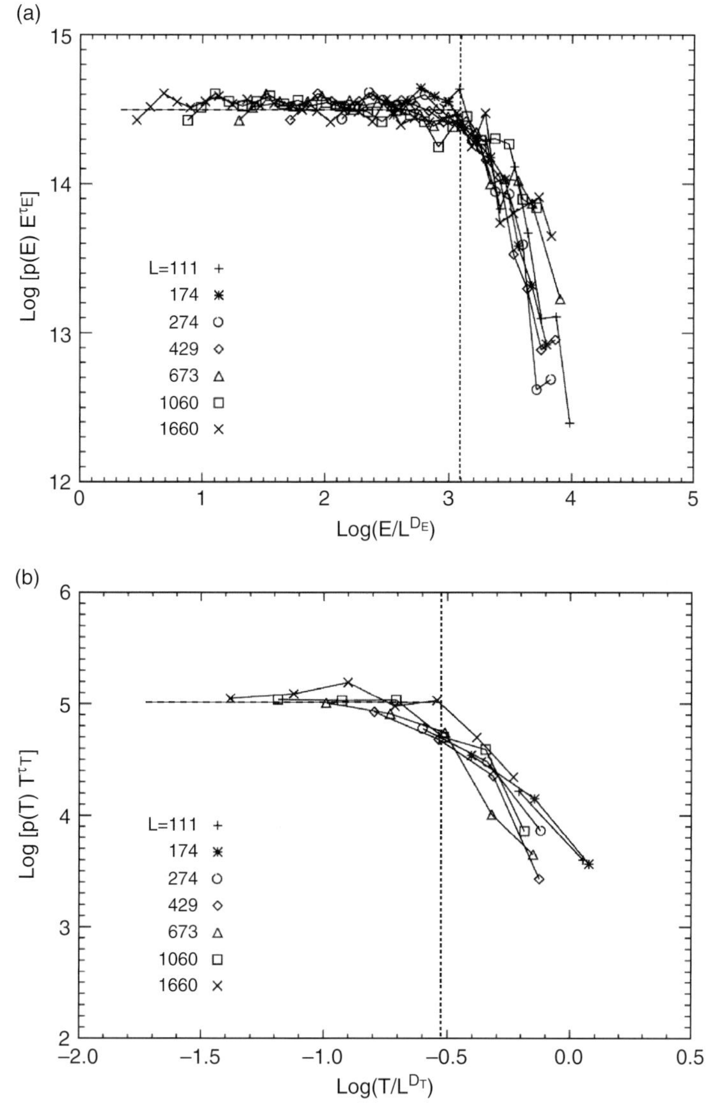

Figure 2.12. Rescaled versions of the probability distributions $P(E)$ and $P(T)$ of auroral emission events of energy E and lifetime T in arbitrary units confirming their data collapsing and cutoff properties. (From Uritsky et al. 2006.)

were obtained for solar maximum and minimum. Because of the large difference in average emission during solar min and max, the threshold w_a relative to $\langle w \rangle$ for each data set was defined separately.

Figure 2.13 shows the probability distributions and their statistics fitted approximately by power laws: $p(T) \sim T^{-\tau_T}$, $p(E) \sim E^{-\tau_E}$, and $p(A) \sim p^{-\tau_A}$. The exponents were nearly independent of the choice of w_a. The results strongly support the hypothesis that the corona operates in a dynamic critical state. We shall come back in later chapters and discuss these results in connection with the concept of multifractals.

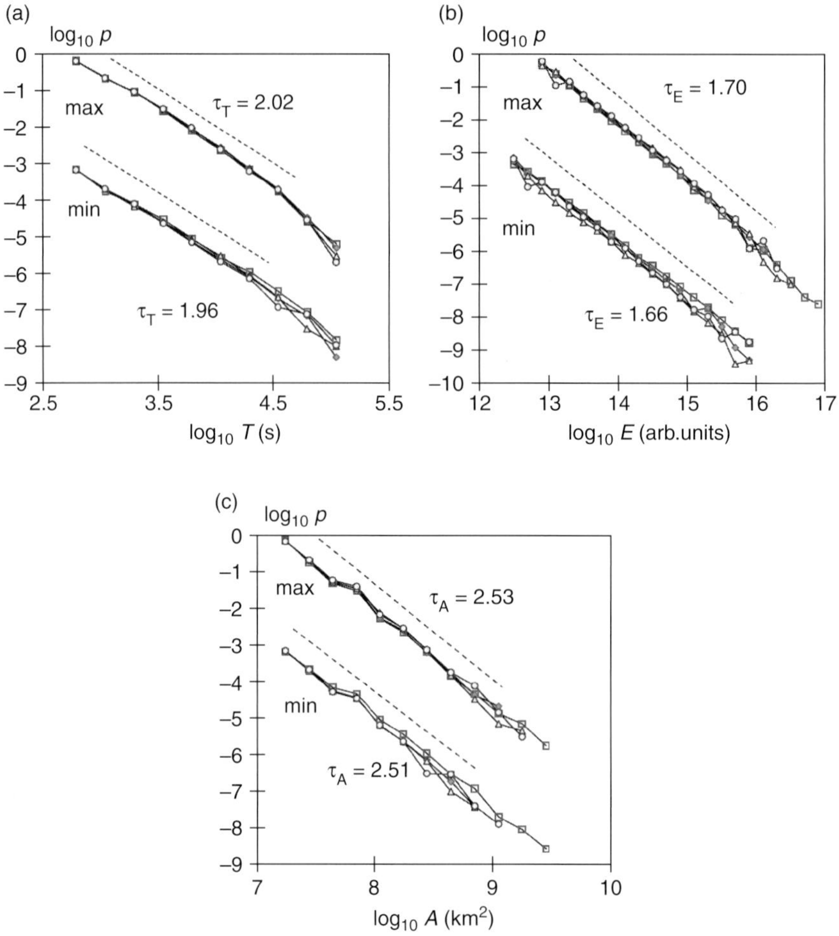

Figure 2.13. Probability distributions of spatiotemporal coronal avalanche lifetime T, EUV emission flux E, and peak area A, for solar max and min at different thresholds w_a (with different geometrical shapes). The solar min distributions are shifted downward for comparison. (From Uritsky et al. 2007.) (For a color version of this figure, please see the color-plate section.)

2.4.4 Modeling Substorms by Running Sandpiles

Uritsky and colleagues (2001a; and references therein) employed a two-dimensional running sandpile to model substorms. Specifically, they considered a 2D square lattice of linear dimension L characterized by a site-potential energy Z expressed in arbitrary units (analogous to the local slope of sandpiles). The system is repeatedly disturbed from outside by adding some amount of energy S to randomly chosen elements at each time step. The element at the grid point (x, y) was considered stable if $Z(x, y) \leq Z_{th}$ where Z_{th} was an excitation threshold. Otherwise, the element became unstable and lost four units of energy that were transmitted to its nearest neighbors. In addition, it was assumed that during the process of energy transfer to each neighbor there was a small amount of energy dissipation equal to $Z_d / 4$. Open boundary conditions were used such that the excess energy ε_0 reaching the edges of the lattice was removed from the system. Therefore, the system was nonconservative even if there were no local dissipation during the energy transfers. Thus, the rules for the cellular automata were:

$$Z_{t+1}(x,y) = Z_t(x,y) - 4 \tag{2.16a}$$

$$Z_{t+1}(x \pm 1, y \pm 1) = Z_t(x \pm 1, y \pm 1) + (1 - Z_d / 4) \tag{2.16b}$$

To characterize the avalanche dynamics and compare it with the magnetospheric disturbances, the total number of unstable sites $F(t)$ was evaluated at each time step t as follows:

$$F(t) = \sum\nolimits_{(x,y)} H(Z_t(x,y) - Z_{th}) \tag{2.17}$$

where H denotes the Heaviside step function and the summation is taken over all pairs of x and y.

Analogy of the preceding cellular automaton and the global behavior of the magnetosphere are as follows. By taking the 2D lattice, the plasma sheet current layer of the magnetotail is assumed to be planar. The magnetic field energy of the current sheet is represented by the site potential energy Z and the plasma instability threshold by Z_{th}. The global geomagnetic perturbation is then represented by $F(t)$.

It was demonstrated that for appropriately chosen small driving term S, the threshold of excitation Z_{th}, and low dissipation level Z_d, the dynamics of the model could reach a stationary state of criticality. As discussed in Section 2.3, the susceptibility for uniform driving for sandpiles is given by the relation

$$\chi = \lim_{h \to 0} \chi_{h,\varepsilon} = \partial \rho_1(h) / \partial h \tag{2.18}$$

where $\rho_1(h)$ is the density of unstable (i.e., active) sites. That is, $\rho_1 = \langle F \rangle / L^2$ with $\langle\ \rangle$ denoting a time average. The simulated result indicates that the susceptibility scales with the dissipation rate as $\chi \sim \varepsilon^{-1}$ over a broad range of ε, which agrees with the one-site mean field theory for small driving rate as shown in Eq. (2.8), Figure 2.14.

To simulate a substorm event in the simulation, the stationary state of the system such as that previously mentioned was disturbed by the sudden increase of the driving term S, analogous to the southward reversal of the interplanetary magnetic field (IMF). This causes the total number of unstable sites $F(t)$ to increase. Then, the excitation threshold was reduced (mimicking the triggering of the substorm due to some current driven instability). This caused increased bursts of activity or jumps of $F(t)$, which corresponded to the breakup and active phase of the substorm with increased dissipation (supercriticality). Afterward, the system began its recovery phase and restored most of its lost energy (subcriticality). By the meticulous choices

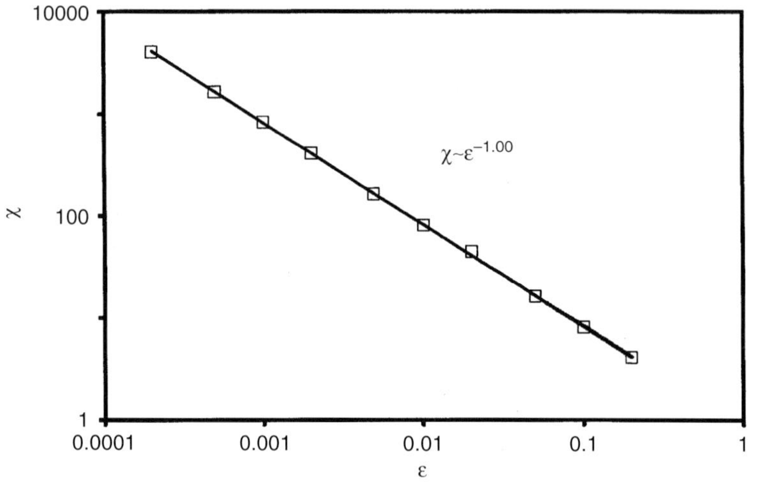

Figure 2.14. Susceptibility as a function of the dissipation rate. The line corresponds to the mean field theory $\chi \sim \varepsilon^{-1}$ of Section. 2.3 for the sandpile model. (From Uritsky et al. 2001a.)

of the parameters, Uritsky and colleagues were able to simulate the variations of the *F(t)* similar to an observed time series of the AE index, Figure 2.15. And, the power spectra of *F(t)* exhibited the distinct features of broker power law similar to that of the observed AE(t) series, Figure 2.16, demonstrating the Hwa and Kardar (1992) running sandpile characteristics.

2.4.5 Revised Forest-Fire Model

In 1990, Bak and colleagues suggested an interesting forest-fire model that supposedly contained the basic ingredients of a dissipative critical system. It was a cellular automaton model defined on a d-dimensional lattice whose sites were either empty or occupied by green or burning trees. The system was updated as follows: at an empty site, a tree could grow with the probability p and a green tree could become a burning tree if one of its neighbors was burning. Also, a burning tree would leave the site as an empty site. However, it was soon realized that the system did not exhibit the expected critical behavior for small values of p (Grassberger and Kantz 1991; Mossner et al. 1992). Instead the system became deterministic with regular spiral-shaped fire fronts.

With a rather clever modification of the model by introducing a small lightning parameter (probability) $f \ll p < 1$ to allow a tree without burning nearest neighbor to catch fire, moderate-sized simulations seemed to demonstrate the desired dissipative dynamic critical behavior for $\theta = f / p \to 0$ and $p \to 0$ (Drossel and Schwabl 1992). Further studies of much-larger-scale simulations (Grassberger 2002; Pruessner and Jensen 2002; 2004), however, indicated such apparent critical behavior might actually be transient states and the system actually exhibited

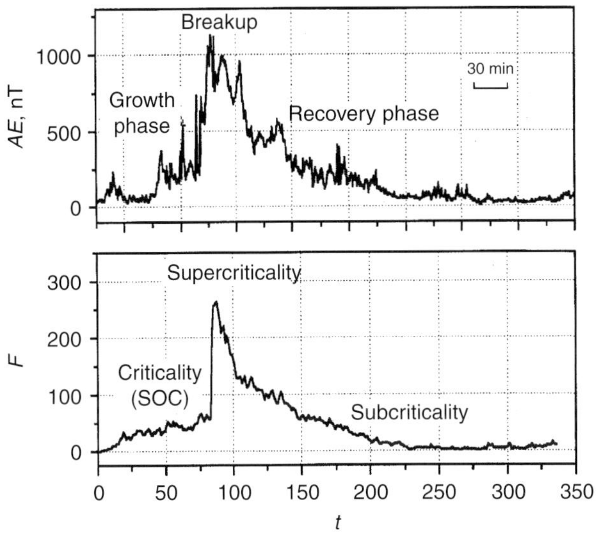

Figure 2.15. Illustration showing the similarity between the evolution of a typical magnetospheric storm on July 2, 1978 onset time 2:30 UT in terms of the observed AE series (top) and the development of the nonstationary perturbation $F(t)$ of the sandpile model caused by first the increase of the driving term S and then the decrease of the threshold Z_{th} (bottom). (From Uritsky et al. 2001a.)

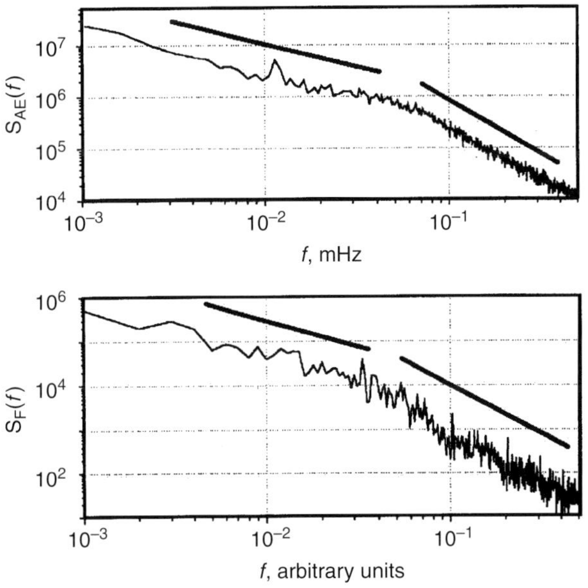

Figure 2.16. Power spectra of the observed AE series of 1978 and the simulated $F(t)$ of the sandpile model of Figure 2.15. Both exhibit the distinct features of broker power-law behavior of running sandpiles. (From Uritsky et al. 2001a.)

much more involved complexity behavior (which we shall address in more detail in Chapter 7). Nevertheless, as we shall see, such transient behavior has its interesting manifestations.

In particular, we consider in this section a variation of this modified forest-fire model (MFFM) suggested by Consolini and De Michelis (2001) to model the occurrence of sporadic localized relaxation of the geomagnetic disturbances in the geotail plasma sheet. Their "revised" model consisted of a 2D square lattice with periodic boundary conditions. The lattice was to mimic a 2D plasma sheet of the magnetotail, and the empty sites represented those regions where plasma conditions were stable, the green trees represented locally unstable regions, and the burning trees were associated with the relaxation (local dissipation) sites.

For this revised forest-fire model (RFFM), the growth probability p_i was driven by a 1D time-stepped coupled map

$$y_{i+1} = (B/2\pi) z_i \sin(2\pi y_i) \tag{2.19a}$$

$$z_{i+1} = C z_i (1 - z_i) \tag{2.19b}$$

that provided a chaotic signal that had approximately a power-law spectrum.

Specifically, they chose

$$p_i = \alpha + \beta y_{i-1} \tag{2.20}$$

The purpose of this choice was to use this chaotic signal to model the statistical driving of the solar wind upon the magnetosphere.

The lightning probability f_i, however, was expressed as a function of the density of green trees ρ^t and was time dependent. Thus it was assumed that there was a coupling between the local dynamics and the density of the local unstable sites. Specifically,

$$f_i = \gamma \exp(\delta \rho^t_{z-1}) \tag{2.21}$$

In these expressions, $\alpha, \beta, \gamma, \delta$ were parametric constants and were assigned the values: $\alpha = 0.0005$, $\beta = 0.15$, $\gamma = 1.0 \times 10^{-4}$, $\delta = 6.0$ for the cellular automaton calculation of a 120^2 lattice. These parametric values ensured that

$$\langle f \rangle / \langle p \rangle < 1 \tag{2.22}$$

to inhibit the occurrence of fire spreading all over the lattice, that is, to prevent the possibility of percolation paths of green trees that interconnect the whole lattice.

Figure 2.17 represents a sequence of six snapshots of the time evolution of the cellular automaton for the RFFM. The intermittent nature of the relaxation process is vividly evident and very similar to the localized reconnection processes in the plasma sheet of the Earth's magnetotail. To quantify the statistical features of this simulated intermittency, the global activity as represented by the number of burning

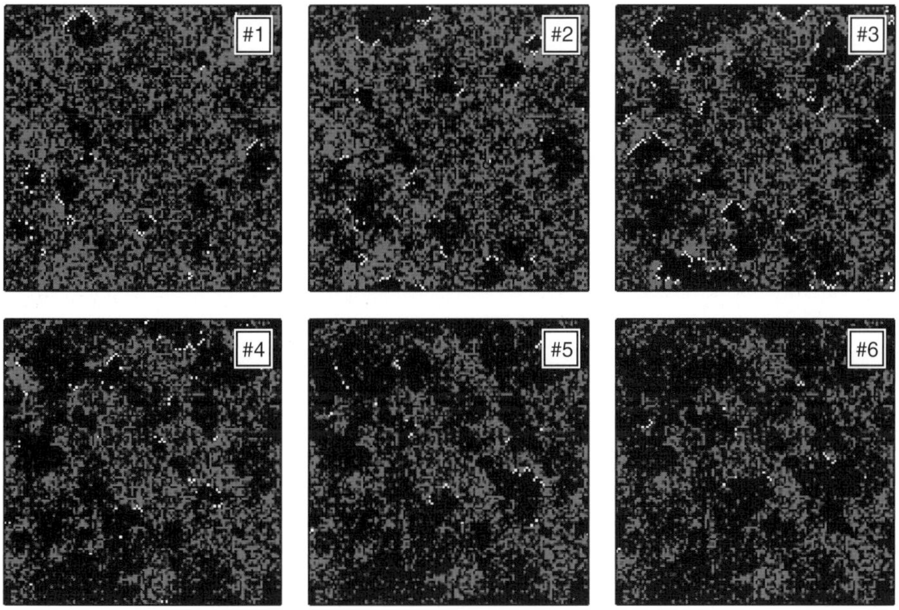

Figure 2.17. Sequence of snapshots of the RFFM cellular automaton. Time interval between snapshots is 10 timesteps. Stable sites (black background), unstable sites (diffused patches), and relaxing sites (sparsely distributed sharp spots). Note the localized and intermittent character of the relaxation processes. (From Consolini and De Michelis 2001.) (For a color version of this figure, please see the color-plate section.)

trees n^f and the probabilities of the size s and lifetime T defined as follows were studied:

$$s = \int [n^f(t) - \hat{n}^f] H[n^f(t) - \hat{n}^f] dt \tag{2.23}$$

$$T = \int H[n^f(t) - \hat{n}^f(t)] dt \tag{2.24}$$

where H was the Heaviside step function and $\hat{n}^f$ a suitably chosen activity threshold.

Figure 2.18 displays log-log plots of the probabilities for a typical cellular automaton-calculation with lattice size (120×120) and $\hat{n}^f = 33$. Straight-line fits indicate power-law behavior over a wide range of scales, similar to those observed for AE indices (Consolini 1997; Takalo 1993). Power spectrum density of the global activity as displayed in Figure 2.19 display the distinct feature for running cellular automata models; namely a high-frequency power-law exponent of near 2.0 and a low-frequency power-law exponent near 1.0 similar to those observed for the AE index (Tsurutani et al. 1990), and running sandpiles (Consolini 1997; Hwa and Kardar 1992; Uritsky et al. 2001). The low-frequency exponent indicates cooperative behavior due to the interference of fire events.

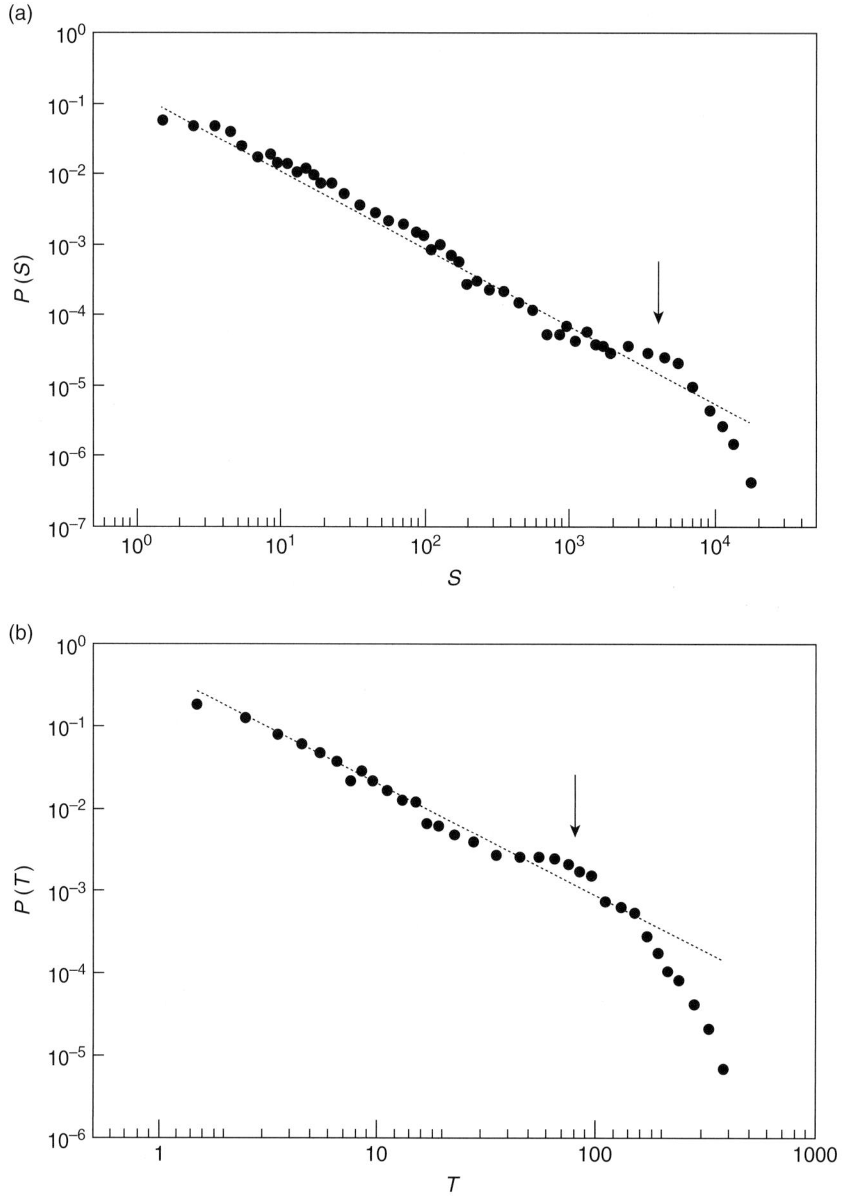

Figure 2.18. Probability distributions functions of the activity burst size s (panel a) and lifetime T (panel b) obtained from the RFFM simulation. Dashed lines are best fits with exponents of (-1.15 ± 0.05) and (-1.36 ± 0.03), respectively. Beyond the two arrows in the panels the probabilities indicate possible finite-size effects. (From Consolini and De Michelis 2001.)

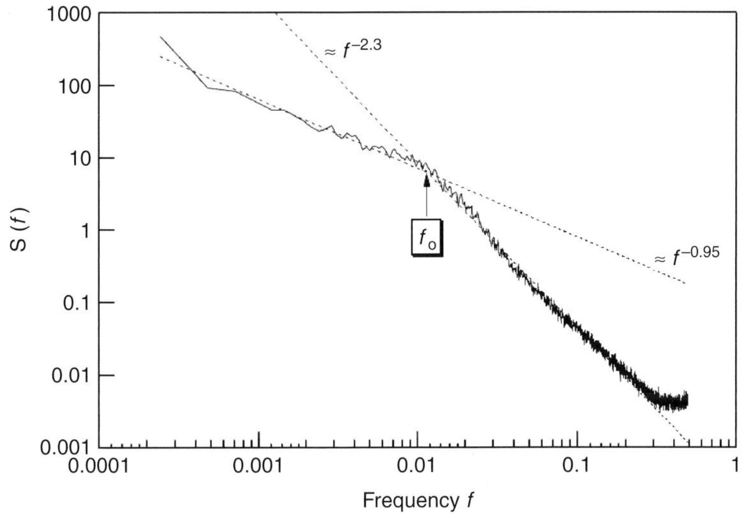

Figure 2.19. Power spectrum $S(f)$ of the global activity of RFFM (sum of relaxation sites as a function of time). The broken power law exhibits signatures of the observed AE index of Tsurutani et al. (1990). (From Consolini and De Michelis 2001.)

2.4.6 Running Coupled Maps Applied to Global Magnetotail Dynamics

In the previous subsections, running sandpile and revised forest fire models were discussed. These cellular automaton models identified the minimum set of rules that might produce the complexity of the global dynamics of the space plasmas from the point of view of dynamical criticality such as SOC. To introduce some basic physics into the formulation of magnetotail complexity, Takalo and colleagues (1999; 2001) considered several related continuously driven coupled map lattice models based on the principles of electricity and magnetism in moving media. For example, they considered a magnetic field model based on the Faraday's law for a moving conducting fluid element:

$$\partial \mathbf{B} / \partial t = -\nabla \times (\eta \mathbf{J}) + \nabla \times (\mathbf{v} \times \mathbf{B}) \tag{2.25}$$

where **B** and **E** are the magnetic and electric fields, **J** the current density, **v** the velocity of the fluid element, and η is the anomalous resistivity. In the absence of other dynamic equations for the plasma medium in this simple approximate formulation, the authors set the second term of Eq. (2.25) as an externally prescribed source term $S(\mathbf{r},t)$ that depended on space and time.

To characterize the current sheet in the magnetotail, the lattice was chosen to be planar in the (x, y) plane with a magnetic field of magnitude $B(x, y)$ normal to it. Thus, with $\mu_0 \mathbf{J} = \nabla \times \mathbf{B}$ and $\nabla \bullet \mathbf{B} = 0$, Eq. (2.25) reduced to a driven diffusion equation for the magnetic field as follows:

$$\frac{\partial B}{\partial t} = \frac{1}{\mu_0}\left[\frac{\partial}{\partial x}\left(\eta \frac{\partial B}{\partial x}\right) + \frac{\partial}{\partial y}\left(\eta \frac{\partial B}{\partial y}\right)\right] + S(x, y, t) \tag{2.26}$$

To discretize the preceding equation, the magnetic field was averaged over each cell (across the section of a rectangular flux tube) of the lattice. The first term on the right-hand side of Eq. (2.26) was then discretized as $(\eta / \mu_0)L_{x,y}$ with

$$L_{x,y} = \frac{1}{(\Delta x)^2}\left[\left(B_{x-1,y} + B_{x+1,y} + B_{x,y-1} + B_{x,y+1}\right) - 4B_{x,y}\right] \tag{2.27}$$

The Laplacian $L_{x,y}$ represented a measure of the current inside the flux tube because

$$\oint \mathbf{J}\bullet d\mathbf{l} = -\iint \nabla^2 \mathbf{B}\bullet\mathbf{n}da \tag{2.28}$$

In outer space, the plasma resistivity is negligible except when anomalous dissipation occurs due to turbulence and instabilities such as those induced by enhanced localized currents. In this application, such effects were represented by a step function for the cell resistivity $\eta_{x,y}$:

$$\eta_{x,y} = \eta H(\left|L_{x,y}\right| - L_{cr}) \tag{2.29}$$

where η was the normalized anomalous resistivity (a parameter for the model), H the step function, and L_{cr} the critical current.

For the source or driving term, $S(x,y,t)$, a time series of the southward component of the interplanetary magnetic field multiplied by the solar wind speed was used to randomly and locally perturb the lattice cells. This was quite reasonable as the southward magnetic field creates an open magnetic field allowing the feeding of solar wind energy into the magnetosphere.

Based on the preceding coupled map lattice model, calculations were performed similar to the cellular automaton algorithm of 2D running sandpiles. The shape of the critical state of the magnetic field with zero boundary conditions for a 50×50 grid is shown in Figure 2.20a for $\eta = 0.25$ and $L_{cr} = 0.5$ for relatively slow driving. It appeared that whatever the shape of the state of the magnetic field was used as the initial state, it eventually evolved into this state. After reaching this shape the magnetic field only fluctuated on top of it, Figure 2.20b. These intermittent localized fluctuations (evaluated as currents at the cell boundaries) were then the structures studied in this analysis, Figure 2.21.

Distributions of the durations and sizes of two hundred thousand avalanches were obtained (Figure 2.22). The input of the analysis was made at a 10×10 area on the top of the magnetic field of Figure 2.20a. The duration of an avalanche is calculated as the number of iterations needed to find the next relaxation configuration and the size is calculated as the time integral over all unstable sites during an avalanche. Both distributions exhibited power-law behavior.

The power spectrum of the time series of the square of the diffusion term (redistributed field) in Eq. (2.26) as well as the overflowing field (dissipation) at the boundaries are given in Figure 2.23. Both spectra have slopes of –0.9 at the

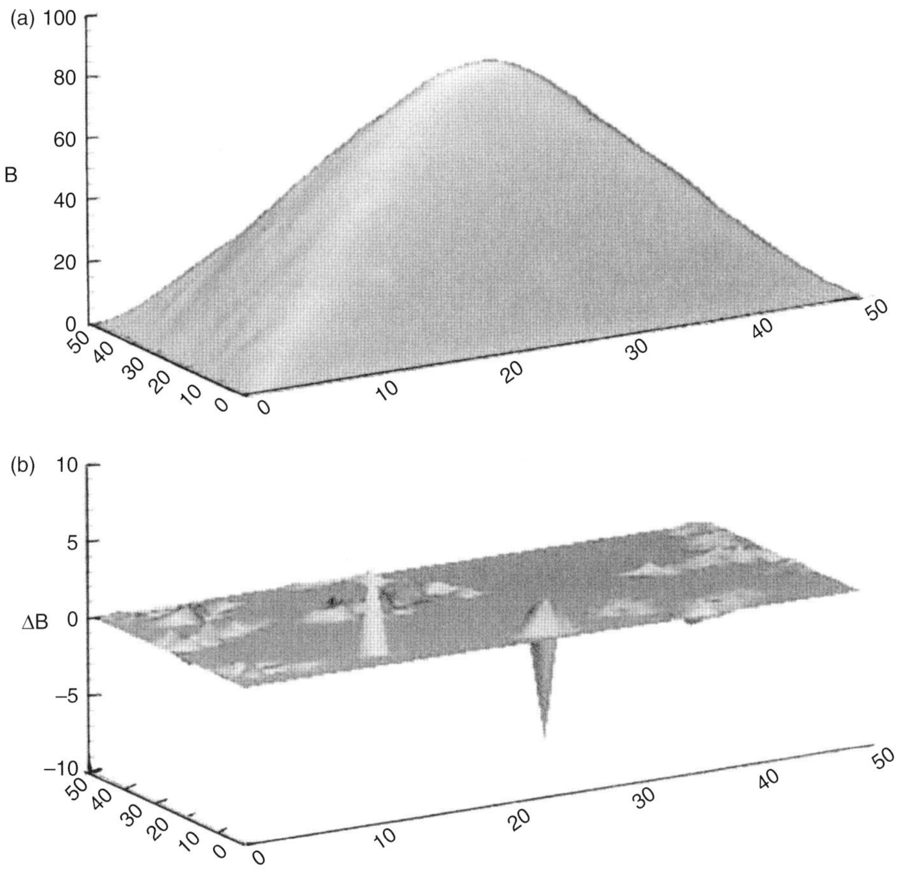

Figure 2.20. (a) Configuration of the magnetic field of the coupled map lattice (CML) magnetotail current sheet model near dynamic criticality. The height of the pile describes the magnetic field intensity B in arbitrary units. (b) Instantaneous time derivative (indicative of the intermittent localized fluctuations) of B between consecutive iterations. (From Takalo et al. 2001.)

low frequency ends. At the high frequency end, a slope of −2.2 for the redistributed energy was observed. The data were not sufficient to ascertain the spectrum slope of the dissipated energy at high frequencies. These signatures are more or less consistent with the characteristics of the running sandpile model of Hwa and Kardar (1992).

2.4.7 The Magnetic Field Reversal Model

Instead of relying on performing complexity studies based on coupled map lattices, investigators were also able to introduce basic physics into their analyses by investigating the dynamical phenomena in space plasmas based on direct numerical simulations of physically motivated models. We select one simple model, the Lu-Klimas fast-field annihilation model of magnetic field reversal, for discussions

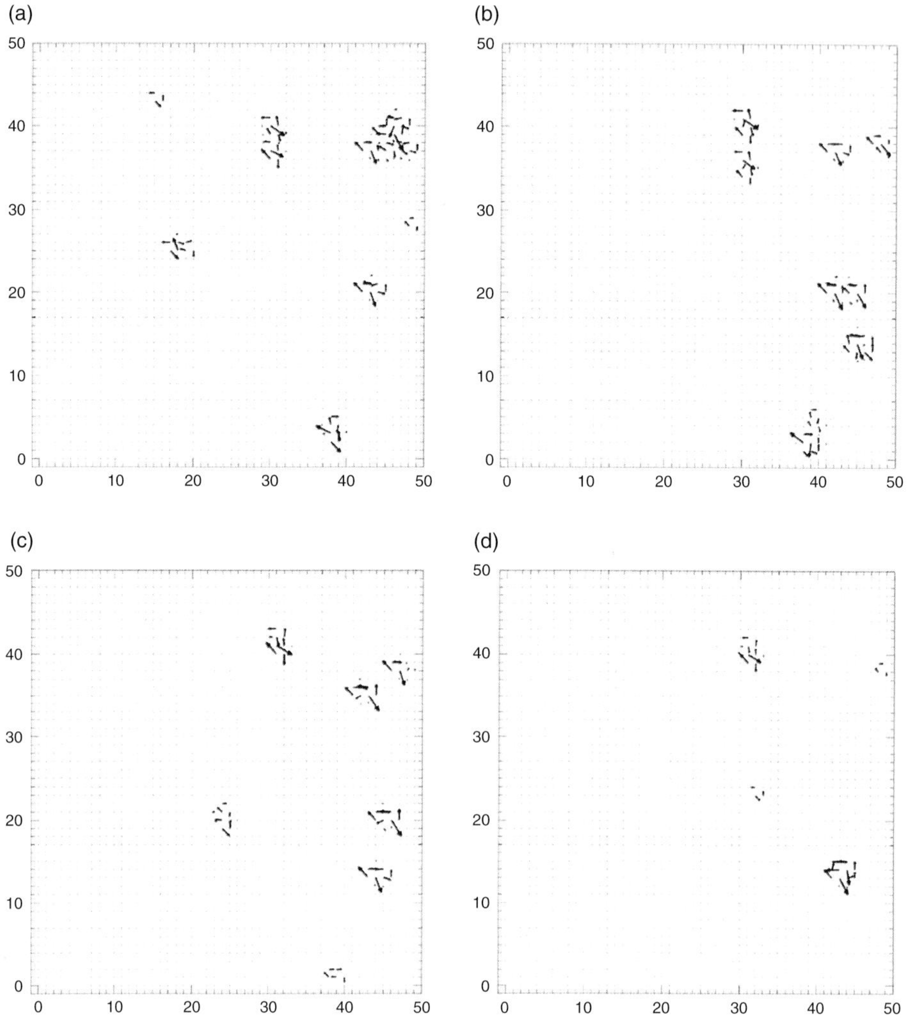

Figure 2.21. Four consecutive snapshots of the localized currents (arrows of different lengths in arbitrary units) around the cells of the CML simulation of Figure 2.20. (From Takalo et al. 2001.)

and illustrations. The model has the basic ingredient of a driven diffusive equation (Klimas et al. 2000):

$$\frac{\partial B_x}{\partial t} = \frac{\partial}{\partial z}\left[D(z,t)\frac{\partial}{\partial z} \right] B_x + S(z,t) \tag{2.30}$$

It may be obtained through the drastic reduction of the resistive MHD (magnetohydrodynamic) system in an idealized one-dimensional limit. In (2.30), $B_x(z,t)$ is the magnetic flux and $S(z,t)$ is a source term that could be interpreted as that characterizes the convection of the magnetic flux, although in discussions it will be arbitrarily prescribed and assumed to be independent of time. The diffusion coefficient $D(z,t)$

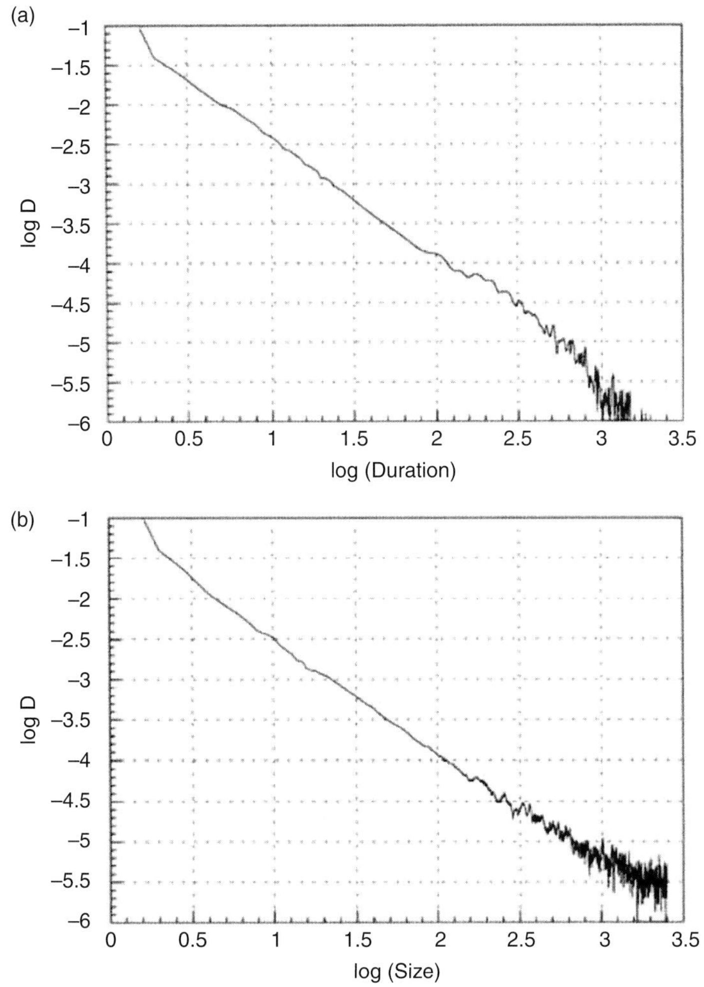

Figure 2.22. (a) Distributions of the durations of two hundred thousand avalanches as defined in the text of the CML simulation of Figure 2.20. (b) Distribution of the sizes of the avalanches for the same simulation. (From Takalo et al. 2001.)

is linearly proportional to an anomalous resistivity with a hysteresis characteristic as stipulated in (2.31):

$$\frac{\partial D}{\partial t} + \frac{D}{\tau} = \frac{Q(|\partial B_x / \partial z|)}{\tau} \tag{2.31}$$

where $Q = D_{\min}$ (low state) for $|\partial B_x / \partial z| < k$, and $Q = D_{\max} \gg D_{\min}$ (high state) for $|\partial B_x / \partial z| > \beta k$ with $\beta < 1$. Thus, Q is double-valued and dependent upon the history for $\beta k < |\partial B_x / \partial z| < k$. The value k may be thought of as a stability threshold that Klimas and colleagues (2000) attributed to some sort of current driven instability (e.g., Lui et al. 1991). The value of Q remains in the low state ($= D_{\min}$) until the threshold is reached and switches to the high state ($= D_{\max}$). It stays at the high state

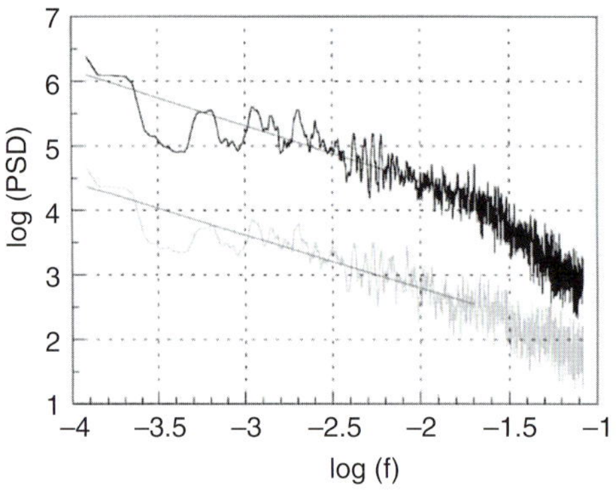

Figure 2.23. PSD of the time series of redistributed energy (upper curve) and dissipated field at boundaries (lower curve) of the CML simulation of Figure 2.20. The scale of log PSD is arbitrary and the graphs are spaced in the vertical direction for clarity. (From Takalo et al. 2001.)

until $|\partial B_x / \partial z| < \beta k$. This anomalous diffusive model is essentially that originally suggested by Lu (1995), which was motivated by his studies related to the phenomena of self-organized criticality (Bak et al. 1987); in particular, those related to the probability distributions of solar flares (Lu 1995; Lu and Hamilton 1991). Unlike the Klimas's model, the source term in Lu's original formulation was assumed to be random and the boundary conditions were chosen to allow transport through the boundaries.

Klimas and colleagues (2000) and Uritsky and colleagues (2001b) have studied their model numerically on a spatial interval $-L \le z \le L$ subject to the boundary conditions $\partial B_x / \partial z = 0$ at $z = \pm L$, that is, no transport of B_x through the boundaries. They considered antisymmetric solutions with a field reversal at $z = 0$ and a time-independent source term $S(z,t) = S_0 \sin(\pi z / 2L)$ that may be considered as a steady convection of B_x into the field reversal region. Setting $\tau = 1$ as the fundamental time scale and $L = 20$, they found that there exist a range of choices of parameters for which the dynamical model evolves in a sequence of quasiperiodic loading $(dE / dt > 0)$ – unloading $(dE / dt < 0)$ cycles where the total field energy $E(t)$ is defined by:

$$E(t) \equiv \int_{-L}^{L} dz B_x^2 (z,t) \tag{2.32}$$

During the loading intervals, when the system is quiet, the slow increase of $E(t)$ is the sole evolutionary feature of the system. As loading continues, the field gradient $|dB_x / dz|$, that is, the current density, reaches the critical level for Q somewhere in the field distribution (usually at a pair of points symmetrically placed near $z = 0$). Subsequently a complicated process of unloading ensues, which is best understood in terms of the spatiotemporal evolution of the diffusion coefficient $D(z,t)$ as depicted in Figure 2.24 during the initial stage of development of the unloading process, where the color scale of the figure indicates the intensity of the diffusion coefficient. In the figure, a white dot is placed at each point where the critical level of $|dB_x / dz|$ – or current density – for Q is reached and $Q(z,t) = D_{\max}$. As the

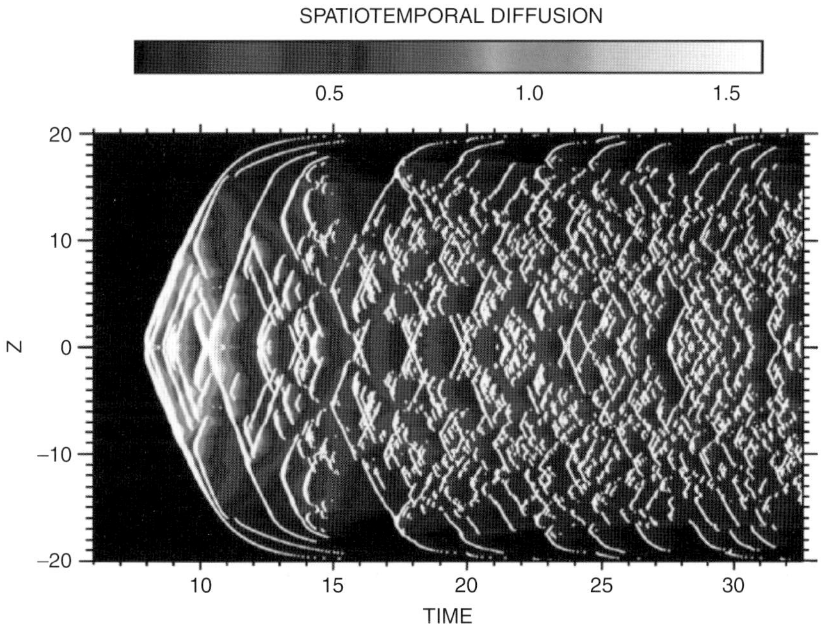

Figure 2.24. Spatiotemporal evolution of the diffusion coefficient $D(z,t)$ in arbitrary units for the Lu-Klimas field reversal model during the initial stage of development of an unloading process. A white dot is placed at each point where the critical level for Q is reached and $Q(z,t) = D_{max}$. (From Uritsky et al. 2001b.) (For a color version of this figure, please see the color-plate section.)

instabilities spread and propagate toward both the center and the edges of the field distribution, more current sheets (i.e., sheet regions with strong current density) are excited and their topology becomes increasingly complex. The current sheets generate diffusion and the diffusion coefficient $D(z,t)$ is at a local maximum just after the passage of a current sheet at any given position. This strong localized diffusion enhances the transport of the magnetic flux toward $z = 0$ where fluxes of opposite signs meet and annihilate. This process of generation of complicated current sheet topology, transport of magnetic flux and annihilation adjusts self-consistently with the imposed source function such that there is a dynamically pseudostationary state. And, the loading-unloading process repeats quasiperiodically after an initial transient.

We now proceed to discuss the invariant properties of the simulated result of this one-dimensional magnetic field reversal and annihilation model. There are at least two sets of statistics that may shed light on the scale-free invariant properties of dynamic criticality for the loading-unloading events that emerge from the numerical calculations of the Lu-Klimas model. First, after an initial transient, the sequence of burstlike unloading events seem to occur by clustering around an average field strength, that is, on average, the increasing strength of the field reversal is more or less balanced by the annihilation of the magnetic field in the current sheet. Thus we

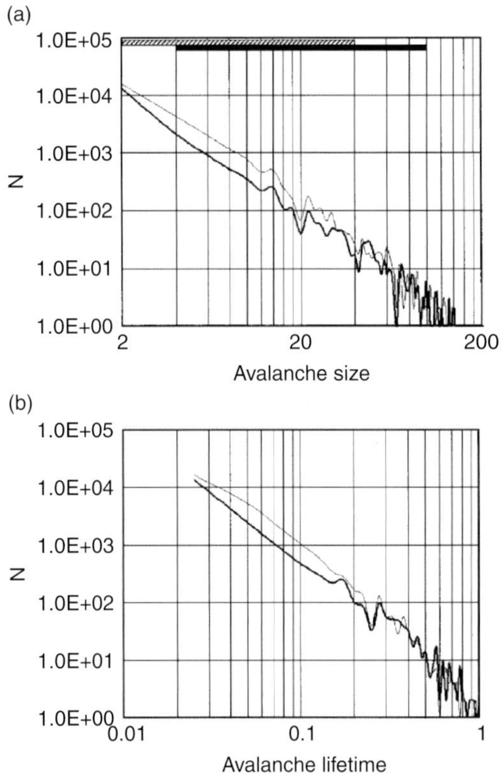

Figure 2.25. Avalanche size (top) and lifetime (bottom) distributions for the Lu-Klimas field reversal model for early (light curves) and late (heavy curves) intervals during a single unloading event. (From Uritsky et al. 2001b.)

expect that the unloading events exhibit power-law distributions in, for example, energy dissipation and burst duration. These results were essentially uncovered by Klimas and colleagues (2000) for the special situation of very low driving rates and $D_{\min} = 0$.

We may also ask the question about the scale-free invariant behavior for the complicated current sheet distributions triggered by the instabilities associated with the anomalous resistivity for a single unloading event. Such small-scale fluctuations again seem to self-consistently balance between the anomalous transport and field annihilation. Thus we may again expect certain universal critical behavior for these small-scale fluctuations. To search for such behavior, Uritsky and colleagues (2001b) defined avalanches in terms of the size and duration of contiguous unstable grid points in the spatiotemporal domain. Power-law statistics of the probability distributions of avalanche size and life time emerged from the statistics of their numerical calculations, Figure 2.25. And these exponents have been demonstrated to satisfy the standard exponent relations known for dynamic critical processes such as those discussed in Section 2.3.

We end this section by mentioning that there are other numerical simulation verifications of criticality behavior from both the 1D and 2D dynamical modeling results, for example, the extension of the preceding work by Klimas and colleagues (2004; 2005) and Uritsky and colleagues (2002b), and the work simulating solar flares by Geogoulis and colleagues (1998), which we shall not discuss here.

More interestingly, Klimas and colleagues (2007) recently demonstrated that the velocity fluctuations of their 2D simulation of the dynamical model exhibited both criticality and multifractal characteristics very similar to those expected for intermittent MHD turbulence, thus, corroborating with some of the basic concepts of interwoven connection and crossover phenomenon between dynamic criticality and intermittent turbulence as espoused in the latter portion of this treatise.

2.4.8 Closing Remarks on Critical Dynamics

In this chapter, we provided a brief review of the phenomenon of critical dynamics with special attention relegated to processes related to the physics of space plasmas; in particular, geomagnetic storms and substorms as well as solar flares. Bak and colleagues' seminal work on SOC (1987) was first introduced and then extended to the study of the occurrence rate of solar flare sizes (Lu and Hamilton 1991). The limitations of the original concept of SOC, such as the requirements of no fine-tuning and nonoverlapping avalanches, were discussed. The suggestion of running sandpiles by Hwa and Kardar (1992) that introduced the concept of overlapping avalanches was then discussed. This idea provided a logical understanding of the nontrivial "$1/f^{\beta}$" flicker noise in nonlinear dynamical systems.

These discussions were followed by an interlude with the introduction of the single-site density field approximation of critical dynamics based on a formulation espoused by Vespignani and Zapperi (1997; 1998). This mean field theory delineated some of the basic concepts of dynamic criticality including that of FSS.

Some observational possibilities of dynamical critical processes in space plasmas were described. Several running cellular automata and a coupled map lattice example were then discussed. These results demonstrated the possibility of the existence of power-law invariants, flicker noise power spectra, and other critical dynamics phenomena in space plasmas. An explicit direct numerical simulation of a field-reversal model was outlined. The results provided a succinct demonstration of the development of critical dynamics of plasmas within magnetized environments, including the verification of some of the expected scaling ideas deduced from the mean field theory.

Most of the examples discussed in this chapter were based on published results performed during the development years of the considered phenomena. Thus the calculations were drawn from rather small scale numerical simulations and system sizes. It is entirely possible that the demonstrated power-law invariants and other scaling properties may change if the system sizes were expanded using more powerful computers and modern numerical schemes. For example, as discussed in Section 2.4.5, the so-called MFFM may actually not be critical in two dimensions even within the general definition of critical dynamics (Grassberger 2002; Jensen and Prussner 2002; 2004). Because the purpose of this treatise is to understand the general phenomena of complexity, we are not really too concerned if a particular numerical or phenomenological model does not exemplify SOC or satisfy the more general definition of critical dynamics. We are more interested in understanding when critical dynamics concepts may hold and/or break down and how such phenomena are interconnected with the more general phenomena of the physics of complexity as shall be demonstrated in the ensuing chapters.

For readers who are interested in delving into more detailed description of SOC-like phenomena in space plasmas, we mention, for example, the interesting results of cellular automata, coupled map calculations, and direct numerical simulations of dynamic models during the developing years of Vlahoes and colleagues (1995), Einaudi and Velli (1999), Georgoulis and colleagues (1998), and Vassilliadis and colleagues (1998) on solar flares. And, of course, there are also the definitive books on SOC by Jenson (1998) and on critical phenomena in natural sciences by Sornette (2000) that cover topics not only on criticality but also of other relevant processes in the more general complexity phenomena to be discussed in the remainder of this monograph and beyond. In addition, there appeared recently a voluminous collection of extended articles on SOC or dynamic criticality systems edited by Aschwanden (2013), which contains a treasure trove of elegant discussions and analyses of the subject.

3

Physics of Complexity

3.1 Introduction

In Chapter 2, we demonstrated that certain space plasma phenomena had signatures of scale-invariant, power-law behavior. And such behavior might perhaps be interpreted as those exhibiting forced and/or self-organized criticality (FSOC/SOC), or, more generally, critical dynamics. A natural follow-up question to raise at this point is: "Are most of the complexity behavior in space plasmas critical dynamics-like, or are there other types of complexity behavior that may permeate through the space plasma environment, particularly within the turbulent domain?" As it has been noted in Chapter 1, there existed phenomena in equilibrium phase transitions, which could be categorized as "multicritical," and the "scaling exponents" for such processes might acquire different values and characteristics as the relevant thermodynamic variables or parameters crossed over (migrated) from one universality class to another. Can such type of phenomena or other phenomena perhaps even more complicated exist in the dynamical complexity processes of space plasmas? Before we proceed to explore such possibilities – that is, to understand why other situations of dynamical complexity may exist and how such states are related to the concepts of FSOC or critical dynamics – it would be prudent to ask exactly how complexity phenomena may arise in the space plasma environments from the physical points of view. It would also be useful to understand the reasons for the existence of such states and their associated varieties of global and local "invariant properties" of complexity from the theoretical point of view.

In the ensuing sections of this chapter, we shall explore the physical reasons of the origin of complexity in space plasmas. We shall also discuss the coarse-grained dissipative processes that allow the coherent (or pseudocoherent) structures to develop, exist, and interact.

3.2 Coherent Structures in Space Plasmas

As it has been indicated, complexity arises in a dynamical system of many nonlinearly interacting parts when multitudes of coherent (or pseudocoherent) structures stochastically form and interact. Plasmas are known for their ability to form numerous types of coherent structures of varied sizes with scales that are generally much larger than those of the constituent particles in the space medium (such as ions, electrons, and neutrals). The primary reason behind this is closely related to the nonlinear and long-range forces responsible for the dynamical couplings among the constituent particles and species interacting within the electromagnetic environment.

Examples of coherent structures abound in the literature on theories and observations of nonlinear space plasma processes. Such structures may appear as convective forms, nonlinear solitary structures, pseudoequilibrium configurations, or other types of spatiotemporal entities. They may be locally generated or convected from elsewhere. Some of these structures may be more stable and long-lived; others may be less stable and relatively short-lived and sometimes partially formed. Generally, such structures are not purely laminar entities; they are likely composed of bundled fluctuations of all frequencies and spatial scales. Due to the nature of the physics of complexity, it would be futile to attempt to assess and/or study the details and stabilities of each of these countless interacting structures or partially formed structures, though some basic understanding of such structures would be helpful in the comprehension of the full stochastic complexity that may result from their underlying nonlinear dynamics. In the following, we shall describe two examples of such coherent structures.

3.3 Alfvénic Resonant Structures

As an example of such coherent structures, let us consider the behavior of Alfvénic resonant structures in magnetized plasmas, which have been observed and/or conjectured to exist in the solar wind, the magnetospheric cusp, the magnetosheath, the plasma sheet within the magnetotail, the solar corona, and elsewhere in the space environment. We shall begin our theoretical discussion of such structures by considering the plasma medium as a single, charge-neutral, electrically conducting fluid moving in and interacting self-consistently with its electromagnetic environment. Because the plasmas in the space environment are generally highly rarefied, we can typically neglect the dissipative effects from particle collisions and assume the plasma is perfectly conducting and inviscid. The most elemental mathematical formulation for magnetized plasmas under such simplifying assumptions is the so-called ideal incompressible magnetohydrodynamics (MHD). The basic equations in Eulerian forms for such a formulation are as follows:

(1) The equation of continuity for an incompressible medium:

$$\nabla \cdot \mathbf{v} = 0, \tag{3.1}$$

where $\mathbf{v}(\mathbf{x}, t)$ is the macroscopic velocity of a fluid element at the spatial location $\mathbf{x}$ and time t.

(2) The equation of motion for an inviscid medium:

$$\rho d\mathbf{v}/dt = \mathbf{j}\times\mathbf{B} - \nabla p, \tag{3.2}$$

where $\rho(\mathbf{x}, t)$ is the density of the medium and assumed to be a constant, $\mathbf{j}(\mathbf{x}, t)$ is the current density, $\mathbf{B}(\mathbf{x}, t)$ is the magnetic field, $p(\mathbf{x}, t)$ is the fluid pressure, $\mathbf{j}\times\mathbf{B}$ is the Lorentz force, and $d/dt = \partial/\partial t + \mathbf{v}\cdot\nabla$ is the convective derivative. We have neglected the influence of the gravitational body force.

(3) The Gauss's law of magnetism:

$$\nabla\cdot\mathbf{B} = 0, \tag{3.3}$$

(4) The Faraday's law of induction:

$$\partial\mathbf{B}/\partial t + \nabla\times\mathbf{E} = 0, \tag{3.4}$$

where $\mathbf{E}(\mathbf{x}, t)$ is the electric field.

(5) The Ampère's law:

$$\mathbf{j} = \nabla\times\mathbf{B}, \tag{3.5}$$

where, for nonrelativistic motion, the displacement current and space charge have been neglected.

(6) The Ohm's law for a perfectly conducting medium:

$$\mathbf{E} + \mathbf{v}\times\mathbf{B} = 0. \tag{3.6}$$

The preceding equations are written in SI units with the vacuum permeability set equal to one.

These equations may be combined to form the following set of equations of motion and induction:

$$\rho d\mathbf{v}/dt = (\mathbf{B}\cdot\nabla)\mathbf{B} + \nabla(p + B^2/2), \tag{3.7}$$

$$d\mathbf{B}/dt = (\mathbf{B}\cdot\nabla)\mathbf{v}. \tag{3.8}$$

Neglecting the effect of the second term on the right-hand side of (3.7) for the moment, standard arguments lead Eqs. (3.7) and (3.8) to a linearized wave equation characterizing the fundamental propagation of small fluctuations of $\mathbf{v}$ and $\mathbf{B}$. This wave equation admits the well-known Alfvén waves, which are transverse waves to the magnetic field **B** and propagate along the magnetic field direction. The phase velocity v_A of the Alfvén wave is $B/\sqrt{\rho}$, that is, proportional to the strength of the magnetic field and inversely proportional to the square root of the density ρ of the plasma. For such waves to propagate, the operator $\mathbf{B}\cdot\nabla$ must not vanish, that is, $\mathbf{B}\cdot\nabla \rightarrow \mathbf{k}\cdot\mathbf{B} = k_{\|} \neq 0$. In other words, the propagation vector **k** must contain a field-aligned component $k_{\|}$.

General Characteristics of the Alfvénic Resonant Structures

An interesting question to pose at this point is: What happens when the parallel component $k_{\|}$ of the propagation vector vanishes locally? The answer: At such a locality, fluctuations cannot propagate and therefore are localized. Such singularities (resonance sites) in the hydromagnetic medium generally have the shape of three-dimensional space curves. Slightly away from these resonance sites, it may be shown that the fluctuations – which will try to propagate away as Alfvén waves – are held back by the background magnetic field and the plasmas medium, thereby forming the so-called Alfvénic coherent (resonant) structures – which are actually domains of bundled stochastic fluctuations.

Let us now consider the geometric properties of such Alfvénic resonant structures. For an ideal MHD system, any physically acceptable magnetic field must satisfy the solenoidal condition (Gauss's law): $\nabla\bullet\mathbf{B} = 0$, and the density of the incompressibility condition: $\nabla\cdot\mathbf{v} = 0$. Also, any variation of the fields away from the initial values must satisfy the constraints (3.4) and (3.6), that is, Faraday's law and Ohm's law for infinite conductivity. These conditions are, of course, reflected in the combined induction equation of (3.8).

We demonstrate that these constraints are equivalent to an infinite set of integral constraints involving the helicity K, such that

$$K = \int_V \mathbf{A}\bullet\mathbf{B}\, dV \tag{3.9}$$

is an invariant for any singly connected volume V enclosed by a flux surface (with $\mathbf{v}\cdot\mathbf{n} = \mathbf{B}\cdot\mathbf{n} = 0$), where $\mathbf{A} = \nabla\times\mathbf{B}$ is the vector potential. (Our discussions here will need to be modified for multiply connected domains. See, for example, Biskamp [1993]). We proceed by evaluating the comoving derivative:

$$\frac{dK}{dt} = \frac{d}{dt}\int_V \mathbf{A}\cdot\mathbf{B}\, dV = \int_V \frac{\partial\mathbf{A}}{\partial t}\cdot\mathbf{B}\, dV + \int_V \mathbf{A}\cdot\frac{\partial\mathbf{B}}{\partial t}\, dV \tag{3.10}$$

The first integral of (3.10) is gauge invariant for $\mathbf{B}\cdot\mathbf{n} = 0$ on the boundary because

$$\int_V \nabla\chi\cdot\mathbf{B}\, dV = \oint_S \chi\mathbf{B}\cdot\mathbf{n}\, dS = 0 \tag{3.11}$$

where χ is an arbitrary gauge. Choosing the gauge such that the scalar potential vanishes leads to $\partial \mathbf{A} / \partial t = -\mathbf{E} = \mathbf{v} \times \mathbf{B}$ because of the Ohm's law for infinite conductivity (3.6). Therefore, the first integral vanishes. The second integral of (3.10) may be rewritten with the aid of the induction equation (3.8):

$$\int_V \mathbf{A} \cdot \nabla \times (\mathbf{v} \times \mathbf{B}) dV = \oint_S [\mathbf{A} \cdot \mathbf{v}\mathbf{B} - \mathbf{A} \cdot \mathbf{B}\mathbf{v}] \cdot \mathbf{n}\, dS \tag{3.12}$$

This integral also vanishes because, as assumed, $\mathbf{v} \cdot \mathbf{n} = \mathbf{B} \cdot \mathbf{n} = 0$ on the flux surface. Q.E.D.

The restrictions ($\mathbf{v} \cdot \mathbf{n} = \mathbf{B} \cdot \mathbf{n} = 0$ and singly connected region) involved in arriving this result, however, are easily violated especially for a turbulent medium as we shall see.

Taylor's Conjecture

We are interested in the situation in which there are domains with stochastic turbulent fluctuations within which field lines mix indistinguishably. Thus, it will be difficult – in fact, unpractical – to discuss the topology of individual field lines or flux surfaces. Furthermore, the individual field lines, if they may be identified, will break and merge frequently and randomly. Thus, the assumptions involved in deriving Eq. (3.12) are not satisfied in general. However, it was suggested by Taylor (1974) that when the volume integral of (3.9) is taken over a stochastic region, the coarse-grain averaged helicity in a relaxed state would be essentially conserved. This conjecture seems to be generally verified in laboratory experiments. As the domain of the stochastic region near the Alfvénic resonance site relaxes to a statistically stationary energy state under the constraint of this conjecture of conservation of coarse-grained helicity, one may demonstrate using the variational principle that such a resonance domain (coherent structure) will be essentially force free, that is, $\mathbf{j} \times \mathbf{B} = 0$, in coarse-grained sense.

We shall first consider the situation in which the plasma energy is dominated by the magnetic energy, $B^2/2$. Our aim is to search for the stationary energy states under the conjectured constraint of coarse-grained helicity conservation. Thus, using the method of the Lagrangian multiplier technique of calculus of variations, we search for the stationary states of the integral:

$$I = \int_V (B^2 + \lambda \mathbf{A} \cdot \mathbf{B})\, dV \tag{3.13}$$

by varying the function $\mathbf{A}$ subject to the boundary condition of $\delta \mathbf{A} = 0$, where λ is the Lagrangian multiplier and is arbitrary. The result is:

$$\nabla \times \mathbf{B} = \lambda \mathbf{B} \tag{3.14}$$

Using the Ampère's law, we obtain the desired result that the stationary energy states of the resonant domain are characterized by the force-free state: $\mathbf{j}\times\mathbf{B}=0$ in coarse-grained sense. If the kinetic energy of the domain cannot be neglected, then the variational principle will also yield the additional requirement of $\mathbf{v}=0$ for the stationary states.

We are, of course, interested in dynamical states that are far from equilibrium. Thus, in visualizing the relaxed states from the point of view of this *Taylor's conjecture*, we shall consider timescales such that "nearly coherent" structures are formed. These structures actually move, mix, and sometimes merge together while immersed in an otherwise turbulently diffusing plasma medium. Such states are generally metastable.

3.4 Alfvénic Flux Tubes

To obtain some physical insight of the geometries of the Alfvénic coherent structures, let us first consider the special situation in the solar wind and make the reasonable assumption that the perturbed magnetic field fluctuations are much smaller than – and essentially transverse to – the mean magnetic field B_0 which will be temporarily assumed to be uniform for the current discussion (Matthaeus et al. 1990). Thus, let us write $\mathbf{B}=(\delta B_x,\delta B_y,B_0)$, where z is in the direction of the mean magnetic field, and (x, y) are orthogonal coordinates normal to z. The force-free condition for constant B_0 and $\nabla\cdot\mathbf{j}=0$ then leads approximately to the scalar condition $\mathbf{B}\cdot\nabla j_z=0$, obtained by taking the z-component of the curl of $\mathbf{j}\times\mathbf{B}=0$. We have, then,

$$B_0\partial j/\partial z=-(\delta B_x\partial/\partial x+\delta B_y\partial/\partial y)j, \tag{3.15}$$

where j is the z-component of the current density. For convenience, let us introduce the magnetic flux function ψ by writing $(-\partial\psi/\partial y,\partial\psi/\partial x)=(\delta B_x,\delta B_y)$ for the perturbed transverse components of the magnetic field in the (x,y) directions such that the Gauss law of magnetism, $\nabla\cdot\mathbf{B}=0$ is automatically satisfied. Then, j and ψ are governed by Eq. (3.15) and the Ampere's law, $\nabla\times\mathbf{B}=\mathbf{j}$.

A simple example of the flux function and axial current density satisfying the preceding conditions would be the class of circularly cylindrical solutions of $\psi(r)$ and $j(r)$ (Chang 2001a, b). Generally, the solutions would be more involved because of the variabilities of the local conditions of the plasma and the three-dimensional geometry. Moreover, the dynamical coherent structures with the inclusion of plasma pressure and other modifying effects (including the electron-inertia terms) would be even more complicated. However, we expect these structures to be in the form of nearly field-aligned flux tubes, Figures 3.1 and 3.2 (Bruno et al. 2001; Chang 2001a, b; Chang et al. 2004; Wu and Chang 2000; 2001; 2005). Existence of Alfvénic flux tubes in the magnetopause and magnetotail have also been suggested by Tetreault (1992 a, b) and Chang (1998; 1999), respectively.

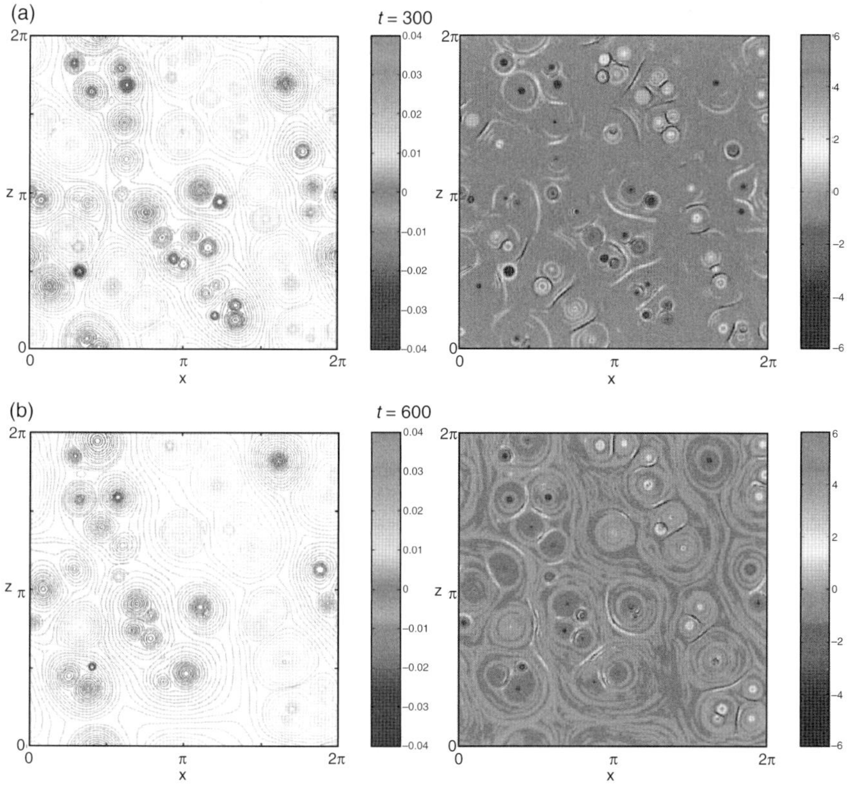

Figure 3.1. Views of interacting coherent structures resulting from a 2D MHD homogeneous turbulence simulation of a periodic square of $(2\pi)^2$ on a $(512)^2$ grid. Left panels are contours of magnetic flux at $t = 300$ and 600. Corresponding current density distributions are given in the right panels. For reference, the sound wave and Alfvén wave traveling times through a distance 2π are about 4.4 and 60, respectively. Shades of gray indicate intensities in arbitrary units. Simulations with a strong uniform magnetic field B_0 will yield similar views in cross-sectional planes normal to B_0. (From Wu and Chang 2001.) (For a color version of this figure, please see the color-plate section.)

Generally, there exist various types of propagation modes (whistler modes, electromagnetic ion cyclotron waves, etc.) in magnetized plasmas. Thus we envision a corresponding number of different types of plasma resonances and associated coherent structures that typically characterize the dynamics of the plasma medium under the influence of an electromagnetic background.

3.5 Alfvénic (Solitary) Vortices

In the previous section, we considered field-aligned flux tubes that are essentially two-dimensional structures normal to the direction of the dominant magnetic field. We also suggested that in the more general setting such structures could have small

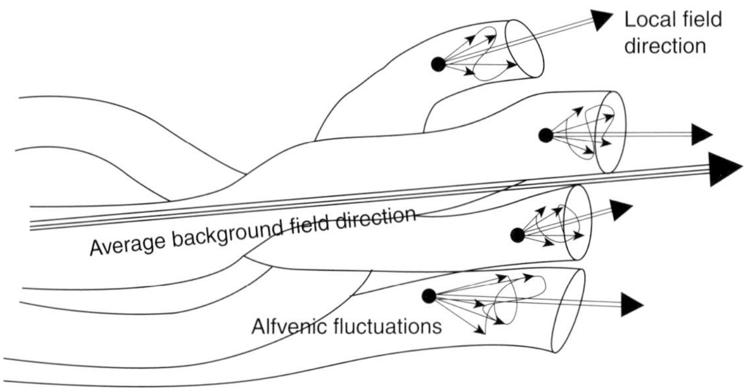

Figure 3.2. Schematics of tangled flux tubes in the solar wind. Each flux tube is characterized by a local magnetic field direction aligned approximately with the background field and the presence of Alfvénic fluctuations makes the magnetic field vector wander randomly about this direction. Moving across the tubes, strong intermittent interactions of predominantly nonpropagating (i.e., resonant) fluctuations are expected. (From Bruno et al. 2001.)

transverse velocities meandering across the field lines as well as small variations along the field line direction. We shall consider a class of such structures that are solitary in nature – that is, structures with particular geometric properties and propagating speeds.

Thus we shall proceed to search for structures that are strongly magnetized in the direction of an axial magnetic field and nearly two-dimensional in the transverse direction with minor axial dependence. We shall base our analysis on the so-called reduced magnetohydrodynamic (RMHD) formulation (Kadomtsev and Pogutse 1974; Strauss 1976). Our discussion follows essentially that described by Biskamp (1993).

The basic assumptions are as follows: (1) The mean axial magnetic field is much larger than the transverse fields with $B_\perp / B_z \sim \varepsilon \ll 1$. (2) The transverse gradients are much larger than the axial variations. (3) The axial magnetic field and velocity are essentially constant. (4) We shall consider only slow time variations. Formally, the consistent ordering of the magnitudes of the various field quantities based upon these assumptions may be summarized as:

$$\begin{aligned} &B_z \sim \nabla_\perp \sim 0(1), \\ &B_\perp \sim v_\perp \sim \partial t \sim \mathbf{v} \cdot \nabla_\perp \sim \partial_z \sim 0(\varepsilon), \\ &p \sim \delta B_z \sim v_z \sim 0(\varepsilon^2). \end{aligned} \tag{3.16}$$

The transverse magnetic field and velocity field may therefore be represented by two scalar fields, the magnetic flux function $\psi\,(x, y)$ and the stream function $\varphi\,(x, y)$ such that

$$B_x = -\partial\psi / \partial y, \quad B_y = \partial\psi / \partial x, \tag{3.17}$$

and, $$v_x = -\partial\varphi / \partial y, \; v_y = \partial\varphi / \partial x. \tag{3.18}$$

We note that the magnetic flux function as defined in (3.17) is equal to the negative of the z-component of the vector potential $\mathbf{A}$ with $\mathbf{B} = \nabla \times \mathbf{A}$.

With these provisions, to the lowest relevant order, the z-component of the curl of the equation of motion (3.2) yields a scalar partial differential equation relating the two scalar functions ψ and φ:

$$d\omega / dt = \{\psi, j\} + B_z \partial j / \partial z, \tag{3.19}$$

where $\omega = \nabla_\perp^2 \varphi$ is the vorticity, $j = \nabla_\perp^2 \psi$ is the axial (z) component of the current density, $\{\psi, j\} = \dfrac{\partial\psi}{\partial x}\dfrac{\partial j}{\partial y} - \dfrac{\partial\psi}{\partial y}\dfrac{\partial j}{\partial x}$ is the Poisson bracket, and $d / dt \equiv \partial / \partial t + \mathbf{v} \cdot \nabla$.

Expressing the magnetic field $\mathbf{B}$ in terms of the vector potential $\mathbf{A}$ in the Faraday's law of induction (3.4) and utilizing the Ohm's law for infinite conductivity (3.6), we have

$$\partial(\nabla \times \mathbf{A}) / \partial t - \nabla \times (\mathbf{v} \times \mathbf{B}) = 0. \tag{3.20}$$

Integrating (3.20), we obtain

$$\partial \mathbf{A} / \partial t = \mathbf{v} \times \mathbf{B} - \nabla \chi, \tag{3.21}$$

where χ is the scalar potential. The z-component of (3.21) gives

$$d\psi / dt = \partial \chi / \partial z. \tag{3.22}$$

According to the ordering of magnitudes prescribed by (3.16), the transverse components of $\partial \mathbf{A} / \partial t$ is negligible comparing to the terms on the right-hand side of (3.21). Therefore,

$$(\mathbf{v} \times \mathbf{B})_\perp = \mathbf{v} \times B_z \mathbf{e}_z = \nabla_\perp \chi, \tag{3.23}$$

or $\mathbf{v}_\perp = \mathbf{e}_z \times \nabla(\chi / B_z)$. Thus the scalar potential is related to the stream function as $\chi = B_z \varphi$ and (3.22) may be rewritten as

$$d\psi / dt = B_z \partial \varphi / \partial z. \tag{3.24}$$

Equations (3.19) and (3.24) are two coupled partial differential equations for the scalar functions ψ and φ. To the lowest order, the only couplings in the z-direction are through the linear gradient terms.

With the z-component of the magnetic field B_z being the only explicit multiplying physical parameter in these equations, they may be conveniently and almost trivially rendered dimensionless. For space plasma applications of a medium containing a specific dominant ion species, besides B_z the additional parameters for such transformations would be the Alfvén speed, the ion gyroradius, and gyrofrequency. (For specific choices, see, e.g., Petviashvili and Pokhtelov 1992; Alexandrova et al. 2006; or Alexandrova 2008.) These equations, then, take on the following forms:

$$d\omega / dt = \{A, J\} - \partial J / \partial z \tag{3.25a}$$

$$dA / dt = -\partial \varphi / \partial z \tag{3.25b}$$

where $A = -\psi$ and $J = \nabla_\perp^2 A$, and all variables may now be viewed as dimensionless.

We now search for special traveling (drift) solitary solutions oriented oblique to the z-axis, which depend upon two independent variables: x and $\eta = y + \alpha z - ut$ where (α, u) are constants. (These constants must be sufficiently small to ensure that the ordering scheme of the RMHD assumption is not violated.) Then, Eqs. (3.25 a, b) become

$$\{\Phi, \nabla_\perp^2 \Phi\} = \{\Psi, J\}, \tag{3.26a}$$

$$\{\Phi, \Psi\} = 0, \tag{3.26b}$$

where, $\Phi = \varphi - ux$, $\Psi = A - \alpha x$.

From (3.26b), we have, $\Psi = f(\Phi)$. Substituting this result into (3.26a) and integrating, we obtain

$$\nabla_\perp^2 \Phi = f'(\Phi) J + g(\Phi), \tag{3.27}$$

which contains two arbitrary functions, f and g. This means that there are nondenumerable admissible drifting solutions that retain their arbitrary shapes as they propagate.

Alfvén Vortices

Among the infinite number of admissible solutions of (3.27), we now consider a special class of solutions with $\varphi = (u / \alpha) A$, the so-called Alfvén vortices, with the current density $J = \nabla_\perp^2 A$ satisfying

$$\nabla_\perp^2 A = 0 \qquad r \geq a \tag{3.28}$$

$$\nabla_\perp^2 A = -k^2 (A - \alpha x - c) \qquad r \leq a \tag{3.29}$$

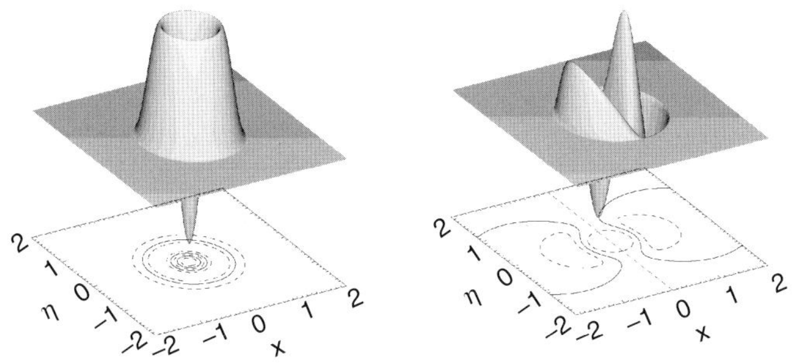

Figure 3.3. (Left) Surface of current density J above the (x, η) plane and contours of potential A (field lines) in the plane for the monopolar Alfvén vortex with radius $a = 1$ and inclination angle $\alpha = 0$. (Right) Surface of current density J above the (x, η) plane and contours of potential A (field lines) in the plane for the bipolar Alfvén vortex with radius $a = 1$, inclination angle $\alpha = 5^{\circ}$ and monopolar amplitude $A_0 = 0$. The general asymmetrical solution is a combination of the two solutions and with $A_0 \neq 0$ (Alexandrova 2008).

where a, k, and c are constants. The solitary solution that falls off at infinity according to a power law and satisfies the matching condition $A - \alpha x = 0$ at $r = a$ so that the current density is continuous has the form (Alexandrova 2008):

$$A = \alpha x (a / r)^2, \qquad r \geq a \qquad (3.30)$$

$$A = A_0[J_0(kr) - J_0(ka)] - \frac{2\alpha x}{kr}\frac{J_1(kr)}{J_0(ka)} + \alpha x, \qquad r < a \qquad (3.31)$$

where J_0 and J_1 are Bessel functions of the zeroth and first order. This solution requires the following dispersion relation:

$$J_1(ka) = 0 \qquad (3.32)$$

The solution consists of a dipole carrier and a monopole core of arbitrary amplitude. From A, the magnetic field fluctuations may be calculated. Typical magnetic field lines of monopolar and bipolar solutions are displayed in Figure 3.3 (Alexandrova et al. 2008).

Such drifting vortices have been detected by Sundkvist and colleagues (2005) in the cusp region and Alexandrova and colleagues (2006) in the magnetosheath region of the Earth's magnetosphere based on measurements from the instruments of the four Cluster spacecraft. In fact, such structures have also been observed in Saturn's magnetosheath with the Cassini mission (Alexandrova and Saur 2008). In reality, however, due to the admissibility of nondenumerable solutions of Eq. (3.27), we envision various magnetic geometries that may arise in space plasmas that exhibit more general shapes of drifting vortexlike Alfvénic structures.

3.6 Coarse-Grained Dissipation and Stochastic Interaction of Coherent Structures

In the previous sections, we discussed plasma coherent structures and provided two specific examples. As it has been alluded to several times before, it is the stochastic process of the dynamic interactions of the coherent structures (or partially formed such structures) that can lead to the complexity behavior of plasma turbulence. Perhaps the single most important attribute of such stochastic interactions is the phenomenon of "coarse-grained dissipation."

Let us demonstrate this idea through the interaction of Alfvénic flux tubes. When coherent Alfvénic flux tubes with the same polarity migrate toward each other, strong local magnetic shear is created, Figure 3.4. Wu and Chang (2000; 2001; 2005) have demonstrated that the existing sporadic nonpropagating fluctuations in the strong local shear region, particularly those close to the neutral sheet (i.e., at the location where the local magnetic field vanishes or more generally where $k_{\parallel} = 0$) will stay in the region and continue to interact nonlinearly, Figure 3.5. However, fluctuations away from the neutral sheet or in the $k_{\parallel} \neq 0$ region are nonresonant and will therefore propagate away along magnetic field lines as Alfvén waves.

Combined with the magnetic shear geometry, the resonant fluctuations will induce nonlinear instabilities near the neutral sheet or $k_{\parallel} = 0$ region. Such nonlinear instabilities, will, in turn, produce more fluctuations – some nonresonant and some resonant. The nonresonant fluctuations will again propagate away as Alfvén waves while the resonant ones will join the other resonant fluctuations and interact nonlinearly, thereby broadening the resonance region, Figure 3.5.

This combined phenomenon of "coarse-grained dissipation" depletes the energy originally contained in the coarse-grained magnetic fields near the shear region, thus, initiating a reconfiguration of the coherent structures. It breaks some of the closed field lines of each of the coherent structures and then reconnects them into

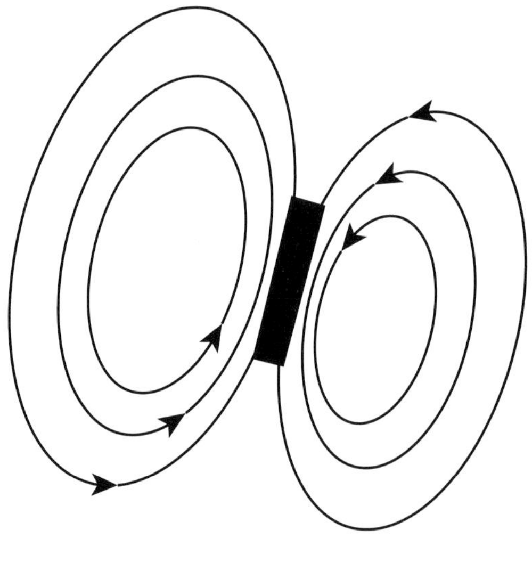

Figure 3.4. Cross-sectional view of interacting coherent structures of the same polarity. Contours indicate constants of magnetic flux and arrows indicate directions of transverse magnetic field. Blackened area is an intense current sheet with strong magnetic shear. (From Chang et al. 2004.)

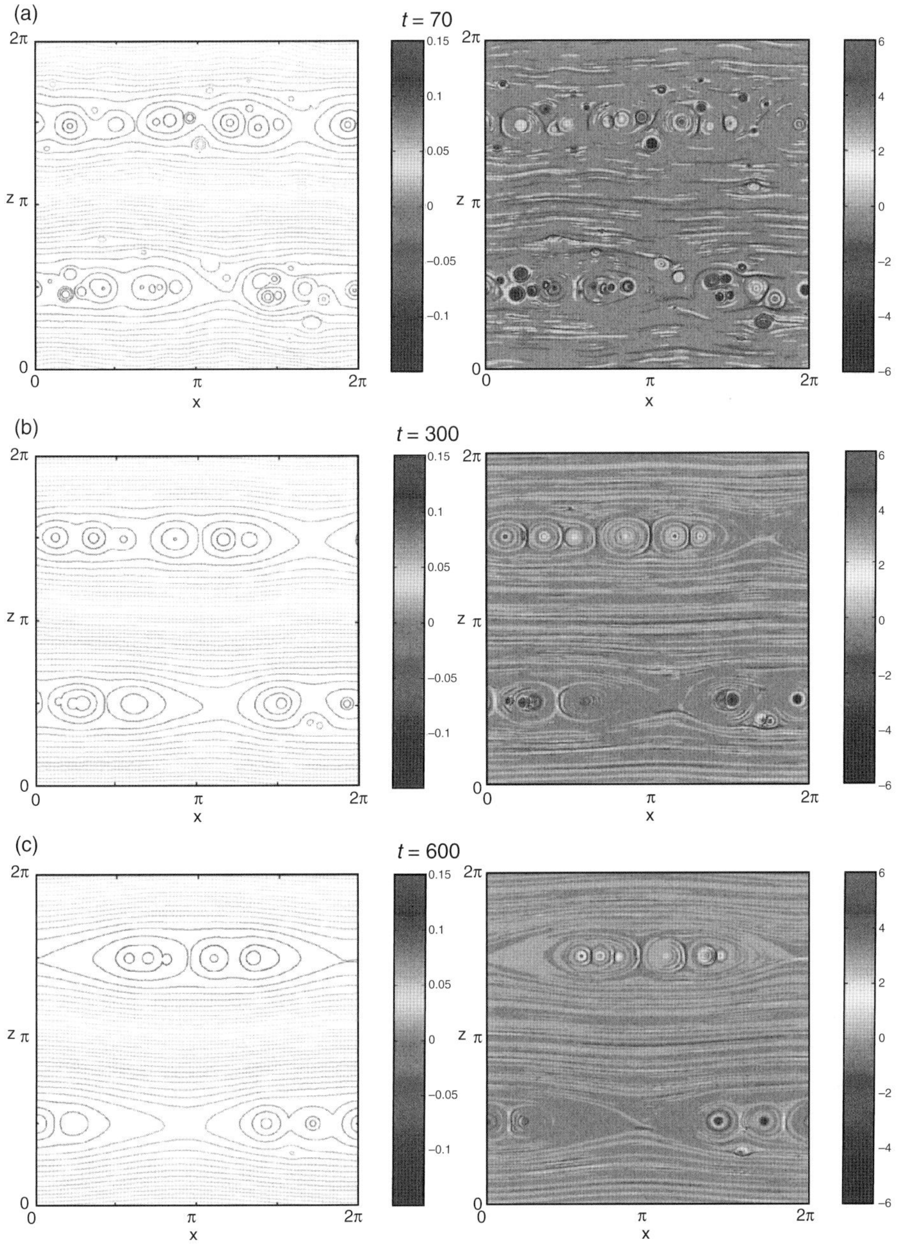

Figure 3.5. 2D MHD inhomogeneous turbulence simulations with magnetic shear of interacting coherent structures on a periodic square of $(2\pi)^2$ of a $(512)^2$ grid. Regions of strong magnetic shear (neutral sheet regions) are located at $z = \pi / 2$ and $3\pi / 2$. Left panels are contours of magnetic flux at $t = 70, 300, 600$. Corresponding current density distributions are given in the right panels. Shades of gray indicate intensities in arbitrary units. For reference, the sound wave and Alfvén wave traveling times through a distance 2π are about 4.4 and 60, respectively. (From Wu and Chang 2001.) (For a color version of this figure, please see the color-plate section.)

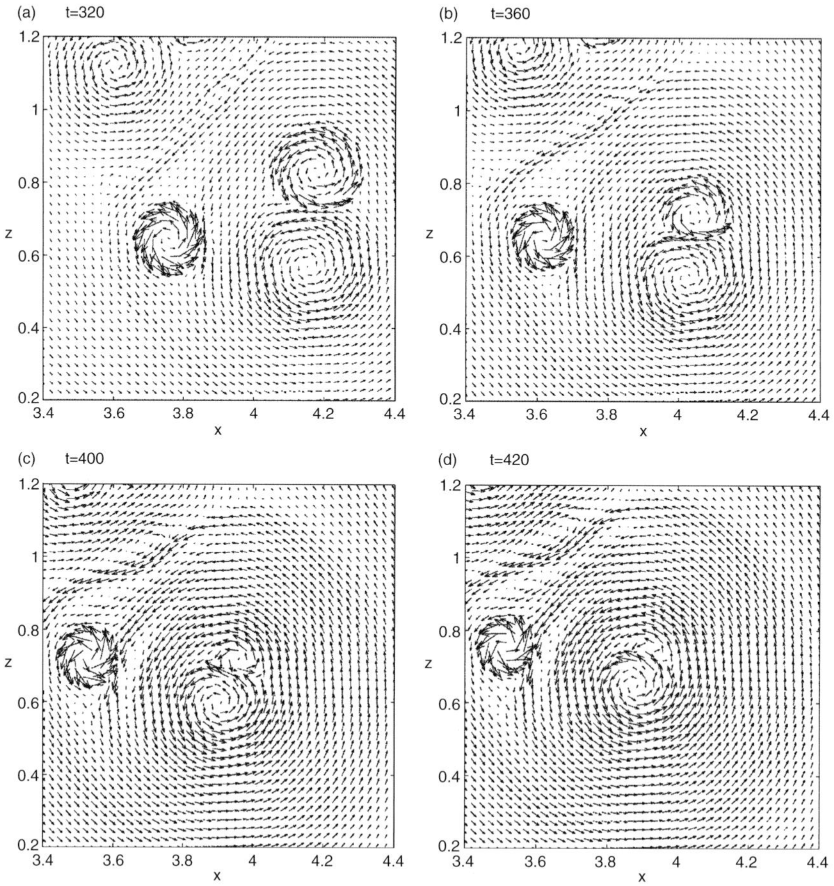

Figure 3.6. Four snapshots of time sequence of a 2D MHD simulation of the dynamic merging of flux tubes. Arrows indicate magnitudes and directions of the magnetic fields. (From Wu and Chang 2001.)

single closed field lines in the coarse-graining averaged sense, Figure 3.6 a, b, c, d. And this process continues with the system adjusting intermittently with the surrounding environment until all available free energies are exhausted, eventually leading to the formation of one single combined coherent structure with one set of concentric closed field lines (again in the averaged coarse-grained sense).

The final state of the resulting coherent structure will have less average energy due to the combined dissipation of Alfvénic propagation of nonresonant fluctuations and nonlinear interactions of the resonant fluctuations (i.e., resonance broadening). Such is the manifestation of the continuous magnetic topological reconfiguration due to the dynamic "fluctuation-induced nonlinear instabilities" under the general constraint of the Taylor conjecture of coarse-grained helicity conservation as suggested by Chang (1998, 1999). And this merging process may repeat over and over again among the coherent structures of the same polarity, from the largest to the smallest scales. We note that this coarse-grained dissipation mechanism can take place without the explicit classical dissipation of viscous interactions or ohmic heating. At smaller scales, of course, kinetic effects, wave-particle interactions, and eventually particle collisions may participate in the overall energy exchange and dissipation.

However, when coherent structures of opposite polarities approach each other due to the forcing of the surrounding plasma, they might repel each other, scatter, or induce magnetically quiescent localized regions.

Under any of the conditions of the preceding interaction scenarios, new fluctuations will be generated. And these new fluctuations can provide new resonance sites, thereby nucleating new coherent structures of varied sizes. Most of the preceding discussions were couched in two-dimensional geometries. Realistic space plasma processes are, of course, generally three-dimensional. To demonstrate some features of the three-dimensional effects, Matthaeus and colleagues (2003) performed a simulation involving both the two-dimensional complexity normal to a strong uniform magnetic field and the field-aligned predominantly Alfvén wave fluctuations. The simulation result (Figure 3.7) displayed some of the intriguingly complex nature of the bimodal MHD fluctuations that are expected in a plasma medium under the influence of a dominant magnetic field (Chang et al. 2004). This kind of tangled geometry of interacting flux tubes leading to dynamical complexity has been described by Bruno and colleagues (2001), who first deduced their existence in the solar wind, as "cooked spaghetti," Figure 3.2. And they are also the origin of the observed sporadic and localized current disruptions in the Earth's magnetotail and elsewhere (Lui 1996; Lui et al. 1988; 1998; 2006), Figure 3.8.

Topological reconfigurations of such nature occur quite frequently for the dynamical interactions of coherent structures in space plasmas and are not limited just to flux tubes; for example, Sundkvist and colleagues (2007) have observed intermittent fluctuations in the magnetosheath due to the reconnection events of multiply nestled current sheets. Similarly, such enhanced intermittency at the interaction regions of the whistler coherent structures has also been surmised by Consolini and Lui (2000) and Consolini and colleagues (2005) in the plasma sheet. All such intermittent interactions are quite akin to the avalanche phenomenon prevalent in sandpile models. They are the origin of the various observed magnetic reconnection signatures in space plasmas.

This stochastic behavior of the interactions of the plasma coherent structures is a phenomenon of "dynamical complexity" as defined in the following text (Chang 2009):

> *"Dynamical complexity" is a phenomenon exhibited by a nonlinearly interacting dynamical system within which multitudes of different sizes of large scale "coherent structures" are formed, resulting in a complicated global nonlinear stochastic behavior for the dynamical system – vastly different from that could be surmised from the original dynamical equations.*

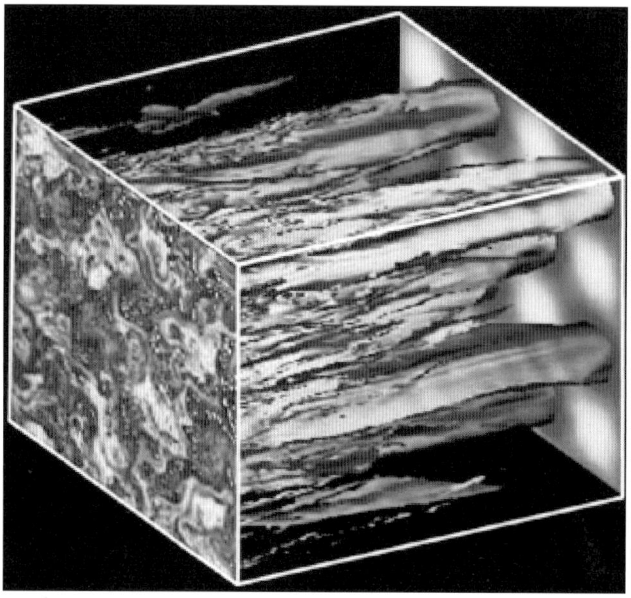

Figure 3.7. MHD simulation of a two-component model, with 80 percent of its energy in two-dimensional modes, exhibiting tangled flux tubes with considerable transverse complexity. (From Matthaeus et al. 2003.) (For a color version of this figure, please see the color-plate section.)

3.7 Complexity and Wavelet Transforms

In Section 3.6, we convinced ourselves that the stochastic interactions of coherent structures of varied sizes generally produce complexity behavior with fluctuations that are localized and sporadic. As we recall from the introductory chapter, one of the more expedient methods to visualize and analyze such chaotic, random fluctuations is through the means of wavelet transforms. Wavelet transform had its birth in signal recognition. In modern days, one of the more popular applications of wavelet transforms is in the area of image processing. Instead of representing a fluctuating pattern or event in terms of a Fourier representation whose basis functions are not localized, wavelet transforms seek to represent an intermittent fluctuating event in terms of the amplitudes and phases of an expansion composed of basis functions that are localized and scale dependent (Farge 1992).

For practical applications, the basis functions of wavelet transforms are usually chosen such that they are square-integrable and generated from a single mother wavelet with its average equal to zero. To be more specific, let us restrict our discussion to one-dimensional events. The basis functions are then generated from a mother function $\psi(x)$ by rescaling and translation, that is, $\psi[(x-b)/a]$, where a represents the scale (dilation or contraction) and b defines the translation along x.

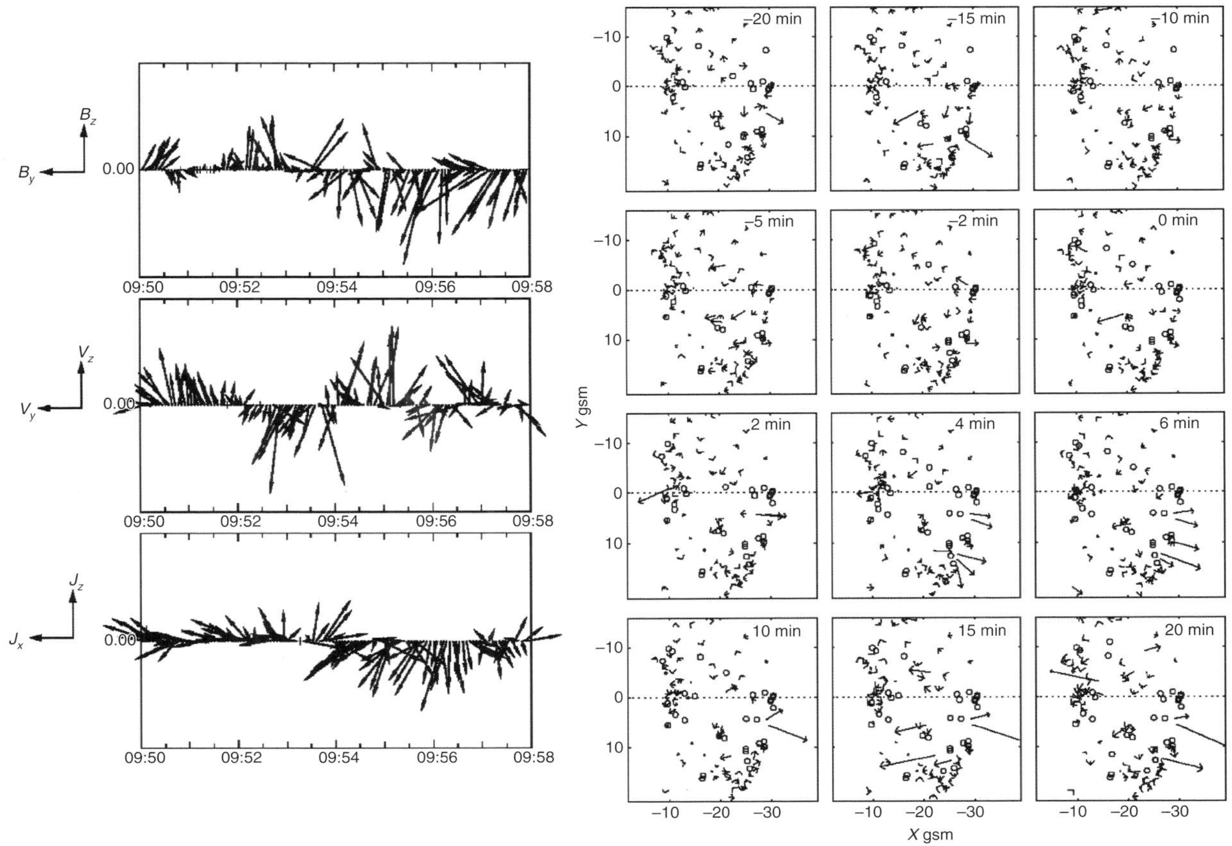

Figure 3.8. Observed sporadic and localized current disruptions in the Earth's magnetotail. (Left) Vector representation of the components of the magnetic field (B_y,B_z), ion plasma flow (V_y,V_z), and current density (J_x,J_z) during a magnetic substorm on August 22, 2001 observed by the Cluster spacecraft at a downstream distance of about nineteen Re (Earth radii) in the magnetotail showing the large variability of these plasma parameters (Lui, private communication). (Right) A temporal sequence of synoptic patterns of total plasma flow showing large variabilities constructed from superposed epoch analysis of 102 magnetic substorm events from Geotail spacecraft observations. The scale is 100 km/s per Re (Earth radius). The acronym GSM stands for Geocentric Solar Magnetospheric coordinate system. The *X*-axis points to the Sun. The *XZ*-plane contains the Earth's dipole axis, and the *Y*-axis completes the right-handed coordinate system. (From Lui et al. 1998.)

To obtain the coefficients of a wavelet transform for a particular fluctuating spatial (or temporal) series $S(x)$, one evaluates the convolution integral as follows:

$$C(a,b) = (K / a)\int S(x')\, \psi[(x' - b) / a]dx' \tag{3.33}$$

where K is a normalization constant. The square of the amplitude of the coefficient $|C(a,b)|^2$ represents the power of the localized signal at $x = b$ with a scale size of $\delta x = a$.

As an example, let us consider the complex-valued one-dimensional Morlet wavelet (1983) defined by the mother wavelet

$$\psi_0(y) = e^{ik\cdot y}e^{-|y|^2/2}, \tag{3.34}$$

a Gaussian enveloped complex exponential wave function with wave vector k. The Morlet wavelet is appealing to the physicists because it fits in with their usual understanding of wave packets. And due to its continuous nature, the transform is amenable to analytical manipulations. Because the wavelets are complex, the transform provides information both of the amplitudes and the phases of the fluctuations.

To carry out the convolution (3.33) for Morlet wavelets using a computer, it may be conveniently accomplished in Fourier space using the Fast Fourier transform with the appropriate normalization factors (Torrence and Compo 1998). And, if the number of oscillations within each wavelet is chosen such that $k_0 = 6$, then the numerical error introduced by the nonzero mean will generally be less than that of the computer round-off errors (Farge 1992). One minor disadvantage of the Morlet wavelets, $\psi[(y - b) / a]$, is that they are not entirely orthogonal to each other due to the extended tails of the Gaussians.

The power spectrum of a one-dimensional fluctuating series may be displayed in a two-dimensional color plot with, for example, the abscissa representing the locations of the fluctuations, the ordinate representing the scale, and the color representing the power intensity. Figure 3.9 is such a graphical representation of the fluctuating spatial series of the square of the magnetic field at the neutral sheet region obtained from a two-dimensional MHD simulation with a sheared magnetic field geometry using the Morlet transform. We note from the figure that the amplitudes of the fluctuations are strongly localized and intermittent, particularly at small scales, a scenario that is expected according to our discussions in Section 3.2 related to the phenomenon of coarse-grained dissipation involving the interaction of multiscale coherent structures. If we wish to include the phase information, then a colored three-dimensional graphical representation of the transform will be needed.

Another popular wavelet transform was introduced by Haar (1910). The one-dimensional mother wavelet is given by

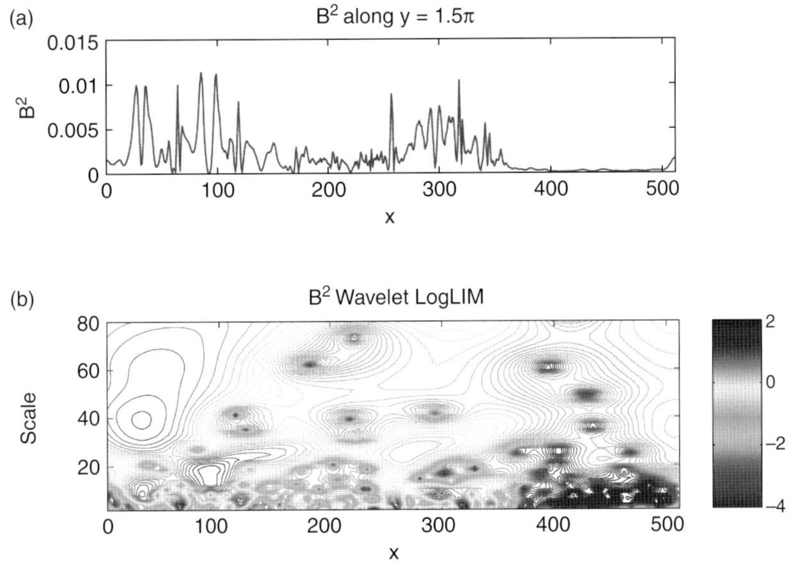

Figure 3.9. *LIM* of the fluctuating spatial series of the square of the magnetic field at the neutral sheet region obtained from a two-dimensional MHD simulation with sheared magnetic field geometry as described in Figure 3.5 using the Morlet transform. Shades of gray indicate intensities in arbitrary units. (From Chang et al. 2004.) (For a color version of this figure, please see the color-plate section.)

$$\psi_0(y) = 1, \text{if } y \in [0, 1/2); -1, \text{if } y \in [1/2, 1), \text{ and } 0, \text{if } y \notin [0, 1). \tag{3.35}$$

The generated wavelets by dilation and translation are real, discrete, and asymmetric. They are fully orthogonal to each other. Because of the discrete and orthogonal properties, Haar wavelets are ideal for the study of events involving discontinuities such as the fluctuations introduced by the intermittent interactions of coherent structures. Haar wavelets are also numerically convenient in practical applications, particularly at higher dimensions.

From the amplitudes, *C(a,b)* of the wavelet transforms, we can calculate the normalized power NP(*a,b*):

$$\mathrm{NP}(a,b) = |C(a,b)|^2 / \langle |C(a,b)^2| \rangle \tag{3.36}$$

where $< \cdots >$ denotes the ensemble average; and, here, the average is essentially the Fourier transformed power for the entirety of the fluctuations. In usual applications, only a truncated finite set of coefficients are used to perform practical calculations. The normalized power is a power spectrum of both scale and location; thus, it is sometimes called the Local Intermittency Measure, or *LIM*. By definition, *LIM* is equal to one for the Fourier transform. Figure 3.10 is a color representation of the *LIM* of the current density using the Haar transform for a two-dimensional homogeneous MHD turbulence simulation. The scale-dependent, sporadic, and localized

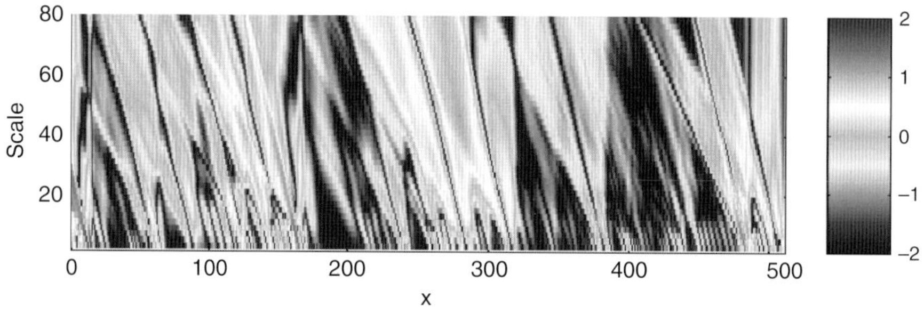

Figure 3.10. *LIM* of the current density J along the x-direction from a two-dimensional homogeneous MHD turbulence simulation as described in Figure 3.1 using the Haar transform. Shades of gray indicate intensities in arbitrary units. (For a color version of this figure, please see the color-plate section.)

powers of the fluctuations are vividly displayed. We note that the intermittency increases as the scale is reduced. And, the intensity can be rather strong at small scales – even when the fluctuations are strongly intermittent and localized.

Normalized power of wavelet transforms have been used to evaluate the *LIM* in the solar wind (e.g., Bruno et al. 1999; 2001), for the AE index (Consolini and Chang 2001; 2002), in the magnetotail (e.g., Consolini et al. 2005; Lui 2002; Vörös et al. 2004), in the cusp (e.g., Echim et al. 2007; Yordanova et al. 2004), and in the auroral zone (Tam et al. 2005).

As it has been pointed out before, intermittent fluctuations generally exhibit non-Gaussian probability distributions especially at small scales. We may look for a quantity that provides a measure of the non-Gaussian nature of the fluctuations as a function of scale. A normalized scale dependent measure related to the fourth-order moment of the coarse-grained probability distribution is called the "Flatness." It indicates whether the data are more peaked or flatter relative to the Gaussian distribution. It is also called the "Kurtosis." It has been suggested by Meneveau (1991) that the average over space (or time) of the square of the normalized power is essentially the scale-dependent Flatness of the probability distribution.

$$\text{Flatness} = \left\langle [\text{NP}(a.b)]^2 \right\rangle \tag{3.37}$$

For Gaussian probability distributions, the value of the Flatness may be shown to be equal to three. For an intermittent event, the Flatness increases without bound at small scales. The rate at which the Flatness increases with the decrease of scale is a measure of intermittency. Figure 3.11 (top) plots the values of the Flatness calculated as a function of scale corresponding to the fluctuations discussed for the magnetic field fluctuations for the 2D MHD simulation that we have discussed previously for given y and fixed t. We note that the value of the Flatness becomes larger than three and continues to increase as the scale becomes smaller and smaller.

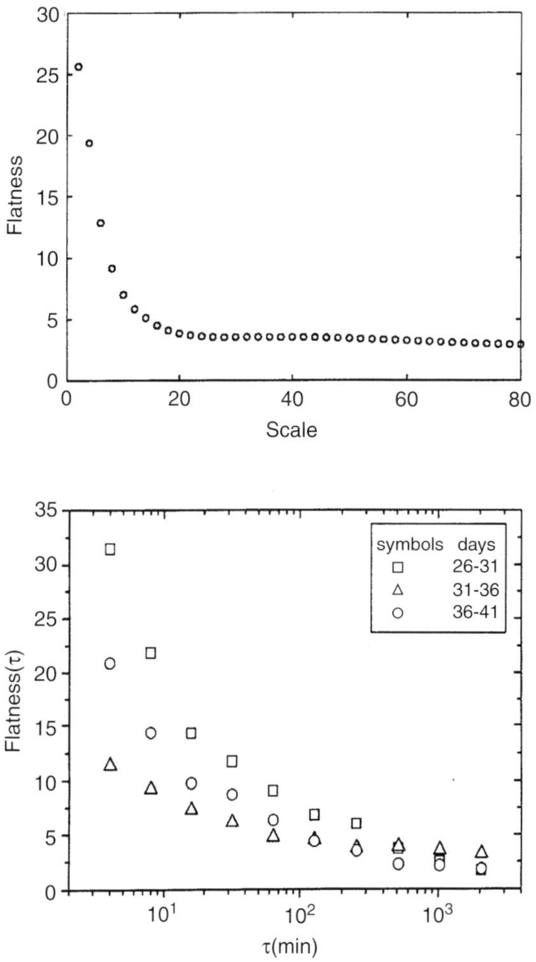

Figure 3.11. (Top) Flatness as a function of scale in units of grid spacing for the intermittent magnetic field fluctuations $B(x)$ calculated with the Haar transform for a 2D homogeneous MHD simulation as described in Figure 3.1 at given time t and for fixed y. (Bottom) Flatness versus time scale for three different time intervals for the solar wind during solar minimum (i.e., when the solar activity is very weak) as measured by the WIND spacecraft. (From Bruno et al. 2007.)

Flatness has been evaluated for the solar wind turbulence by Bruno and colleagues (2007), Figure 3.11 (bottom) based on data observed by WIND (spacecraft launched on November 1, 1994 for magnetospheric and ionospheric space plasma studies). It is one of the tools used by Bruno and colleagues (2001) to arrive at the conclusion that solar wind turbulence is the manifestation of interacting tangled flux tubes or "cooked spaghetti" as depicted in Figure 3.2. Wavelet transforms have been used by various authors to evaluate Flatnesses of intermittent turbulence in other regions of space plasmas (e.g., Echim et al. 2007; Lui 2002; Tam et al. 2005; Vörös et al. 2004; Yordanova et al. 2004). Other more generalized techniques of gauging the intermittency in space plasma turbulence using wavelet transforms have also been employed by Vörös (2000) and Vörös and colleagues (2003; 2005), which we shall not discuss here.

The normalized power and Flatness are only two tools that provide gauges of intermittency of a fluctuating series. One might ask: Are there other wavelet transform techniques that may be employed to characterize the intermittency nature of fluctuating events? The answer is a definite "yes."

For example, in Chapter 5, we shall have the occasion to employ such transforms to study the multifractal characteristics of intermittent fluctuations in the cusp region of the Earth's magnetosphere using the WTMM (wavelet transform modulus maxima) method (Yordanova et al. 2004).

4

Probability Distribution and Structure Functions

4.1 Probability Distribution Functions of Incremental Changes

In our discussions relating to the wavelet transforms, we noted from the displayed examples, that there were noticeable significantly enhanced amplitudes of intermittent fluctuations, particularly at very small scale sizes. Such effects are consistent with our physical interpretation that turbulent fluctuations are generally produced by the interactions and mergings of sporadic and localized coherent structures. And because the coherent structures are numerous and outsized (Figure 3.1), the fluctuations within the narrow interaction regions of these structures (the magnetic shear, current sheet, and resonance overlap regions) would be large and occur relatively more frequently than those that would have been expected from a medium of the original minute plasma particles. One convenient procedure to quantify such intermittent effects is to employ the method of statistics of incremental changes by analyzing the shapes of the probability distribution functions (PDFs) of the changes at different fluctuation scale sizes.

To demonstrate this procedure, let us refer back to the 2D magnetohydrodynamic (MHD) simulations previously described. Consider, for example, the spatial series of the square of the strength of the magnetic field $B^2(x_i)$ for a given time t and at fixed y, with $x_i = i\delta$ where $i = 0,1,2,....,N$ and δ is the grid size of the simulation. To define fluctuations for such a spatial series, we consider the incremental change of $B^2(x_i)$ from a position x_i to an adjacent position $x_i + \Delta$, where $\Delta = k\delta$ with k an integer and form the coarse-grained difference

$$\delta B_i^2(\Delta) = B^2(x_i + \Delta) - B^2(x_i). \tag{4.1}$$

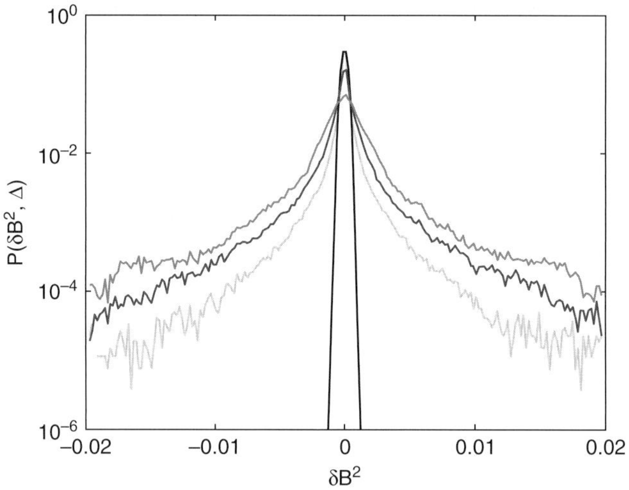

Figure 4.1. Probability distribution function $P(\delta B^2,\Delta)$ of a 2D MHD homogeneous turbulence simulation on a $(512)^2$ periodic lattice for Δ = 2(bottom curve), 4(middle curve), 32(top curve) units of grid spacing. A (black) PDF for Gaussian fluctuations is also shown for comparison. (From Chang et al. 2004.) (For a color version of this figure, please see the color-plate section.)

Assuming statistical homogeneity, we now collect the ensemble of the coarse-grained (incremental) fluctuations as x_i varies over the full range of the simulation box in the x-direction, bin the fluctuations into different sizes, and generate the PDF, $P(\delta B^2,\Delta)$. Figure 4.1 displays the calculated results of $P(\delta B^2,\Delta)$ for such a 2D MHD simulation for the homogeneous case for several coarse-grained scales,Δ. From this figure, we note that the distribution for each scale Δ falls nearly onto a smooth curve with the exception at the tails where scatterings from the mean are more visible. The shapes of the PDF curves deviate from that of a Gaussian distribution especially in the tail regions (heavy tailed). And the deviations of the distributions from the Gaussian become more and more apparent at smaller and smaller scales,Δ. As the scale size increases, however, we expect the PDFs to gradually become scale independent and eventually approach the shape of a Gaussian.

Before we proceed further, let us first make contact with data commonly obtained from in-situ spacecraft observations. Most of the turbulence data in space came from time series measurements obtained by single or multiple spacecraft. And the data are then analyzed based on the Taylor hypothesis (1938), which assumes that the transit time of eddies (bundled fluctuations) is much less than the characteristic evolution time. For fully developed MHD turbulence, the characteristic eddy turnover (evolution) time may be estimated:

$$t(\text{evolution}) \sim (\lambda / 2\pi) / [(\delta B / B_0) v_A] \tag{4.2}$$

where λ is the typical size of the eddy, v_A is the Alfvén speed based on the mean magnetic field B_0 and δB is the average fluctuating magnetic field of the eddies in question. And the transit time of the eddy is:

$$t(\text{transit}) \sim \lambda / v_s \tag{4.3}$$

where v_s is the relative streaming speed between the plasma medium and the spacecraft.

Thus as it has been demonstrated by Matthaeus and Goldstein (1982) the requirement of $t(evolution) \gg t(transit)$ becomes:

$$v_s / v_A \gg 2\pi(\delta B / B_0) \tag{4.4}$$

As pointed out by Tu and Marsch (1995), at one AU for solar wind observations the wind speed (which is essentially the relative speed between the plasma medium and the spacecraft because it is much larger than the spacecraft speed) is of the order of ten times the ambient Alfvén speed and $\delta B / B_0 \approx 0.5$. Therefore, condition (4.4) is satisfied for solar wind turbulence in that region. Because the spacecraft speed(s) is (are) much smaller than the solar wind speed, each time series of data gathered by an instrument aboard a spacecraft essentially samples the spatial fluctuations in the direction of the solar wind, particularly at times when the solar wind turbulence is homogeneous at large scales and fully developed. Thus, for example, a time series of such measurements of the strength of the magnetic field $B(t_i)$, where $t_i = i\delta$ with $i = 0,1,2,....,N$ and δ is the sampling interval, may be interpreted based on the Taylor hypothesis as the spatial series of the ambient turbulent fluctuations with $x = v_s t$ (x: direction of the solar wind velocity). We can then construct the probability distribution function $P(\delta B,\tau)$ or $P(\delta B^2,\tau)$ with the coarse-grained scale $\tau = k\delta$ (k: integer) in analogy to the calculations based on the simulation results discussed previously.

Non-Gaussian probability distribution functions from such statistics have been commonly observed in the solar wind for fluctuations related to the magnetic, velocity, density, and other field variables. See, for example, Burlaga (1993), Feynman and Ruzmaikin (1994), Forman and Burlaga (2003), Hnat and colleagues (2002; 2003), Leubner (2004), Leubner and Vörös (2005), Marsch and Tu (1994), and Sorriso-Valvo and colleagues (1999; 2001). Figure 4.2 depicts the typical PDFs of magnetic field fluctuations obtained from such an analysis for both the slow (~(400 ± 200) km/s) and fast (~(700 ± 100) km/s) solar wind. The similarities between the simulated results and results from such observations as exhibited in Figure 4.2 are quite striking.

Similar observational results have been detected for the magnetic field and velocity fluctuations in the plasma sheet (Angelopolous et al. 1999; Weygand et al. 2005), in the cusp (Echim et al. 2007; Yordanova et al. 2004), for the electric field fluctuations in the auroral zone (Tam et al. 2005; 2010; Tam and Chang 2011), as

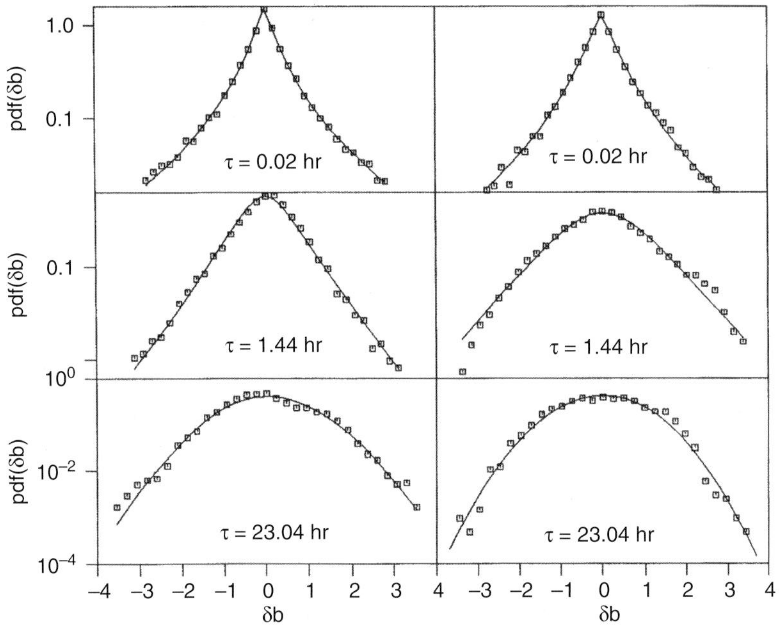

Figure 4.2. PDF for magnetic field fluctuations δb calculated from the observed data obtained by Helios 2 in the (left) fast solar wind (~700 km/s) and (right) slow solar wind (~400 km/s) streams. Solid lines represent fits with the Castaing distribution as discussed in Section 4.2. (From Sorriso-Volvo et al. 2001.)

well as for fluctuations of other field variables, though caution must be exercised in interpreting some of these results as the Taylor hypothesis (or its equivalent) may or may not have been satisfied in the observational regions.

4.2 Castaing Distributions

One popular empirical formula of fitting the shape of the PDFs of the observed intermittent turbulent fluctuations is the Castaing distribution. In 1990, Castaing and colleagues suggested that the intermittent fluctuations might perhaps be viewed as an ensemble of subsets of fluctuations ξ, with each subset having a normal distribution:

$$P_\sigma(\xi) = \frac{1}{\sigma\sqrt{2\pi}} \exp(-\xi^2 / 2\sigma^2) \tag{4.5}$$

where σ is the variance. The intermittency is then assumed to be due to the fluctuations of the variances that satisfy a log-normal distribution:

$$Q_\lambda(\sigma) d\sigma = \frac{1}{\lambda\sqrt{2\pi}} \exp\left(-\frac{\ell n^2(\sigma / \sigma_0)}{2\lambda^2} \right) d\ell n\sigma \tag{4.6}$$

where σ_0 is the most probable variance of ξ, and λ is the variance of $\ell n\sigma$. Combining (4.5) and (4.6) gives the Castaing distribution

$$\Pi_\lambda(\xi) = \frac{1}{2\pi\lambda}\int_0^{+\infty} \exp\left(-\frac{\xi^2}{2\sigma^2}\right)\exp\left(-\frac{\ell n^2(\sigma/\sigma_0)}{2\lambda^2}\right)\frac{d\sigma}{\sigma^2} \tag{4.7}$$

Such fits of data to Eq. (4.7) are shown as solid curves in Figure 4.2 for the solar wind data gathered by Helios 2 (spacecraft launched on January 15, 1976 designed for solar wind plasma and cosmic ray observations) analyzed by Sorriso-Valvo and colleagues (1999; 2001). In general, the fitting parameters, such as σ_0 and λ for the preceding case for symmetrical distributions, depend on the scale of the incremental changes of the distributions. How well does such an ansatz characterize the data depends very much on the type and scale of the fluctuations. Castaing distribution may also be modified to take into account of the skewness of the PDFs. We shall return to this point in Chapter 6.

4.3 Kappa and Other Related Distributions

Kappa distribution was originally suggested by Stanislaw Olbert (Vasyliunas 1968) to describe generalized velocity (v) distributions in kinetic theory of plasmas. It has the form:

$$f(v) \propto \left(1 + \frac{1}{\kappa}\frac{v^2}{v_0^2}\right)^{-(\kappa+1)} \tag{4.8}$$

where v_0 is the most probable speed and κ is a fitting constant. Such a distribution has a near-Gaussian core and a power-law tail. It approaches that of a Gaussian distribution (i.e., without the power-law tail) for very large values of κ, that is, as $\kappa \to \infty$.

Replacing the velocities by the incremental changes of turbulent fluctuations in (4.8), Kappa distributions have been used successfully to describe the PDFs for fully developed turbulence. The power-law nature of such distributions at large values of incremental changes has the characteristics of the statistical distributions of Levy flights (random walks with power-law probability distributions). Truncated Levy flight distributions with exponential cutoffs (due to, e.g., lack of data or finite size effects) are related to the definition of Gamma distributions in statistics.

Though developed originally as an empirical fitting formula, recent developments in the generalized theory of statistical mechanics with nonextensive entropy provide some justification for the Kappa distributions. Classical thermostatistical definition of entropy S (Landau and Lifshitz 1970) in terms of the discrete probabilities of the microstates p_i is:

$$S = -\sum_i p_i \ln p_i \tag{4.9}$$

in units of the Boltzmann constant. Such a definition implies the additive property of entropies. It is commonly named after "Shannon" (1948), the father of information theory, who originally introduced the concept to quantify the expected value of information contained within a message. Extremizing the Shannon entropy (4.9) leads to the Gibbs-Boltmann distribution. For a perfect gas, for example, this process leads to the Maxwell (i.e., Gaussian) velocity distribution in kinetic theory.

A nonextensive definition of entropy (Daróczy 1970) generalizing the entropy of (4.9) has been popularized by Tsallis (1988) as follows:

$$S_q = \frac{1}{q-1}\left[1 - \sum_i p_i^q\right] \tag{4.10}$$

where q, the entropy index, is a real number. Such a definition provides a specific pseudoadditivity (nonextensitivity) property of entropies:

$$S_q(A+B) = S_q(A) + S_q(B) + (1-q)S_q(A)S_q(B) \tag{4.11}$$

where A and B are two independent systems in the sense that the probabilities of A+B factorize into those of A and of B.

For $q \to 1$, expression (4.10) reverts to (4.9). And, when extremized (Beck 2000) under the norm constraint of $\sum_i p_i = 1$ and the energy constraint of $\sum_i P_i \varepsilon_i = U_q$, where ε_i are the energy levels of the microstates and

$$P_i = p_i^q / \sum_i p_i^q \tag{4.12}$$

denotes the so-called escort distributions (Beck and Schlögl 1993), we obtain

$$p_i = \frac{1}{Z_q}(1-(1-q)\beta\varepsilon_i)^{1/(1-q)} \tag{4.13}$$

with $Z_q = \sum_i (1-(1-q)\beta\varepsilon_i)^{1/(1-q)}$.

The β plays the equivalent role of the inverse temperature in kinetic theory. And in the limit $q \to 1$, expression (4.13) reverts to the classical Gibbs-Boltzmann distribution. The introduction of the escort probabilities in place of the original microprobabilities avoids some undesirable divergences of the obtained distributions in multifractal systems (Tanaka 2010; Tsallis et al. 1998).

Beck (2000) and Arimitsu and Arimitsu (2000a, b; 2001) were the first to apply (4.13) to turbulence by suggesting the identification of the energy levels ε_i with the kinetic energy $(\delta\bar{u})^2/2$ (for incompressible flow) of the incremental velocity fluctuations, $\delta\bar{u}$. This is a reasonable procedure for stationary fully developed turbulence based on the assumption that the stationary state of the turbulence on the average maintains pseudostochastic equilibrium among the fluctuations within a given scale. And, therefore, the index q is generally scale dependent. We note that in this context, the existence of the normalization constant or partition function limits $1 \leq q < 3$.

With this identification and the transformation of $(1-q)^{-1} = \kappa$, (4.13) takes on the form of the Kappa distribution (Leubner 2002). In fact, it was suggested that by (ignoring the constraint of the value of q as indicated) and extending the range of κ from $-\infty$ to $+\infty$, a "by-Kappa" distribution may be constructed that sometimes seemed to provide better fittings for the observed intermittent solar wind fluctuation data (Leubner and Vörös 2005).

4.4 One-Parameter Scaling for Self-Similar PDFs

If the (observed or simulated) scale-dependent PDF $P(X,\Delta)$ of incremental changes X exhibit self-similar behavior, then it may be demonstrated that they satisfy the one-parameter scaling form:

$$P(X,\Delta)\Delta^s = P_s(X / \Delta^s) \tag{4.14}$$

where s is the parameter (or scaling exponent) and Δ is the incremental scale. That is, all PDFs essentially collapse onto one master scaling function P_s.

We demonstrate this in the following text by using the invariant and generalized homogeneous function arguments that we employed to derive the finite-size scaling relations described in Chapters 1 and 2. Specifically, we assume

$$P \to P' = e^{\lambda_p \ell} P, \quad X \to X' = e^{\lambda_X \ell} X, \quad \Delta \to \Delta' = e^{\lambda_\Delta \ell} \Delta \tag{4.15}$$

where ℓ is the scaling parameter and the $\lambda's$ are scaling powers. Now, if we further assume that the functional form $P(X,\Delta)$ remains the same under the scale transformations (self-similarity), we obtain the generalized homogeneous scaling function

$$P(e^{\lambda_X \ell} X, e^{\lambda_\Delta} \Delta) = e^{\lambda_P \ell} P(X,\Delta) \tag{4.16}$$

which (with the normalization condition of the PDFs) immediately leads to the expression (4.14) with

$$s = -\lambda_P / \lambda_\Delta = \lambda_X / \lambda_\Delta \tag{4.17}$$

Such scaling behavior seemed to be approximately satisfied for the WIND solar wind data examined by Hnat and colleagues (2002) with $s \approx 0.42$ (Figure 4.3), although closer examination reveals some discrepancies – particularly in the tail regions. Similar result was obtained by Chang and colleagues (2004) for their 2D MHD simulation with $s \approx 0.335$, Figure 4.4. We shall consider the more subtle nature of multifractal (i.e., non-self-similar) scaling properties of more general shaped PDFs in Chapter 6.

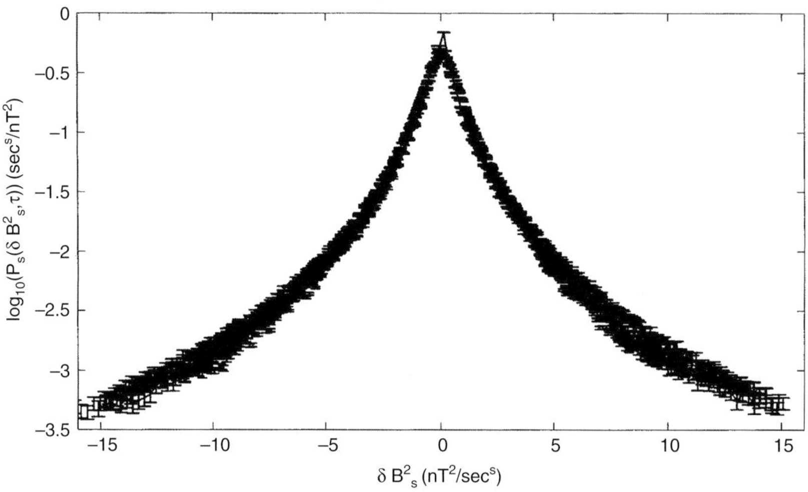

Figure 4.3. Scaled PDF based on the one-parameter scaling form of Eq. (4.14) with $s \approx 0.42$ for the fluctuations of the magnetic field energy density δB^2 as seen by the WIND spacecraft. The data shown correspond to τ between forty-six seconds and twenty-six hours. $\delta B_s^2 = \delta B^2 / \tau^s$. (Hnat et al. 2002.)

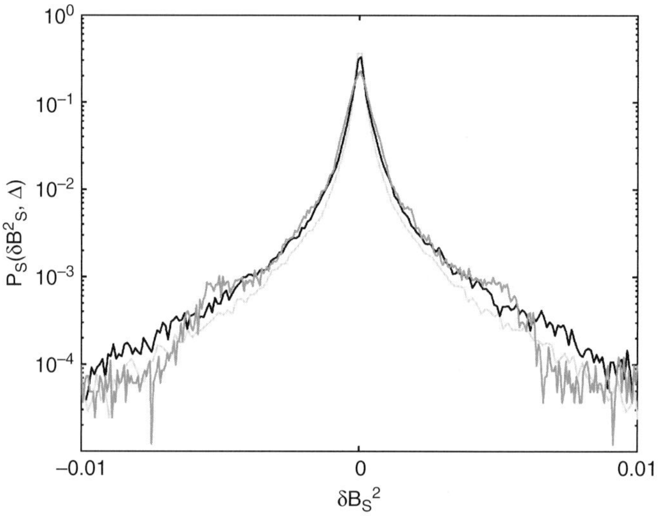

Figure 4.4. Scaled PDF for magnetic field energy density fluctuations δB^2 based on the one-parameter scaling form of Eq. (4.14) with $s = 0.335$ from a 2D MHD homogeneous turbulence simulation on a $(512)^2$ periodic lattice for $\Delta = 2, 4, 32$ units of grid spacing. $\delta B_s^2 = \delta B^2 / \Delta^s$. (From Chang et al. 2004.) (For a color version of this figure, please see the color-plate section.)

4.5 Fractals, Multifractals, and Structure Functions

4.5.1 Fractals

A popular modus operandi designed to study the phenomenon of intermittency is based on the concept of fractals and multifractals. As we have seen, strong turbulence in space plasmas primarily encompass fluctuations generated by the interactions of coherent structures of all varieties and sizes. For illustrative purposes, let us visualize such fluctuations to possess certain geometrical properties in a three-dimensional Euclidean space. Because of their sporadic and localized nature, it is easy to imagine that they generally cannot fill the full three-dimensional space that they occupy at a given time. Or said in another way, the space these fluctuations occupy is only a fraction of the full three-dimensional space. Such geometrical property was popularized by Mandelbrot (1977) when he coined the word *fractals* or fractal geometry.

Actual fluctuations in plasma turbulence generally do not have the conventional geometrical properties. We must then devise some abstract "measure" to quantify the properties of the fluctuations and evaluate its fractal characteristics that may be interpreted with geometrical analogs. Consider, for example, the spatial series of the simulated fluctuations of the strength of the magnetic field, $B(x_i)$, along some constant value of y at time t for the two-dimensional homogeneous MHD turbulence discussed in the previous sections, where $x_i = i\delta$ with $i = 0,1,2,....,N$ and δ is the grid spacing. From this series, we can construct a spatial series by considering, for example, the absolute value of the fluctuations due to the coarse-grained difference of the strength of the magnetic field between two spatial values $x_i + \Delta$ and x_i with $\Delta = k\delta$, some multiple of δ:

$$\delta B_i = |B(x_i + \Delta) - B(x_i)| \tag{4.18}$$

within the spatial interval $X = N\delta$. The simulation is "statistically homogeneous" over the interval X. Thus, we may calculate the ensemble average of δB_i over this interval,

$$S_1(\delta B, \Delta) = \langle |B(x_i + \Delta) - B(x_i)| \rangle \tag{4.19}$$

and use it as a "measure" for the coarse-grained fluctuations of the simulated spatial series. (The choice of taking the ensemble average of the absolute values of the coarse-grained differences instead of the values of the raw differences is for the purpose of better statistical convergence [Biskamp 2003; Vahnstein et al. 1994]. We shall come back to this point again.) We can then plot $\log S_1(\delta B, \Delta)$ against $\log \Delta$ for different choices of Δ (or k). If the result approximates a straight line for some range of Δ for sufficiently small Δ, we can then represent $S_1(\delta B, \Delta)$

as a power of Δ with an exponent ζ_1 within this range. This exponent may be considered as an analog to the classical concept of dimension. It is generally an irrational number and usually cannot be surmised from simple dimensional analysis arguments; thus it is sort of a "fractal dimension" for the particular choice of "measure" described in the preceding text in the limited range where the power law is approximately satisfied.

4.5.2 Structure Functions

It is obvious that if another choice of measure is made, the corresponding fractal number for the same spatial series for small Δ will generally be different. For example, one might wish to look at higher order moments of the PDFs $P(\delta B, \Delta)$ of δB, the so-called structure functions:

$$S_q(\delta B, \Delta) = \int (\delta B)^q P(\delta B, \Delta) d(\delta B) = \left\langle \left| B(x_i + \Delta) - B(x_i) \right|^q \right\rangle \tag{4.20}$$

The motivation here is that different moments emphasize different peaks in the fluctuating series.

Generally, corresponding to each S_q there will be a fractal (structure function) exponent ζ_q for some limited range of small Δ. (In the following, we shall generally omit the mentioning of the limitation, "the results for some range of small Δ," until we reach a point where this limitation becomes very relevant in the discussion.) If $\zeta_q = \zeta_1 q$, then the fractal property of the fluctuating series is fully characterized by the value of ζ_1 (which is usually called the Hurst exponent). Such fluctuations are then said to be "monofractal" or "self-similar," that is, the fractal characteristics for all the moment orders are similar to each other. For general intermittent turbulence, however, ζ_q would be a nonlinear function of q, or "multifractal." It may be easily demonstrated that PDFs that satisfy the one-parameter scaling form of (4.14) obey the monofractal property of $\zeta_q = \zeta_1 q$ with $s = \zeta_1$.

In practical calculations, it has been suggested by Benzi and colleagues (1993) for hydrodynamic turbulence studies that the correlations among the structure functions seemed to be much more apparent than those exhibited by the correlations of individual structure functions directly with the coarse-grained scales. This is particularly true for structure functions based on the absolute values of the differences of fluctuations as it is defined here (Grossmann et al. 1997). This also seemed to be true for plasma turbulence studies (Carbone et al. 1995; 1996; Müller and Biskamp 2000; Politano et al. 1998). Thus, it has become very popular to first determine the relative values of the structure function exponents with respect to the structure function exponent of a particular order, and then determine the exponent for that particular order more accurately by some phenomenological, dimensional, or empirical reasoning. This is similar to the common procedure for critical phenomena in finite systems, where all the scaling properties

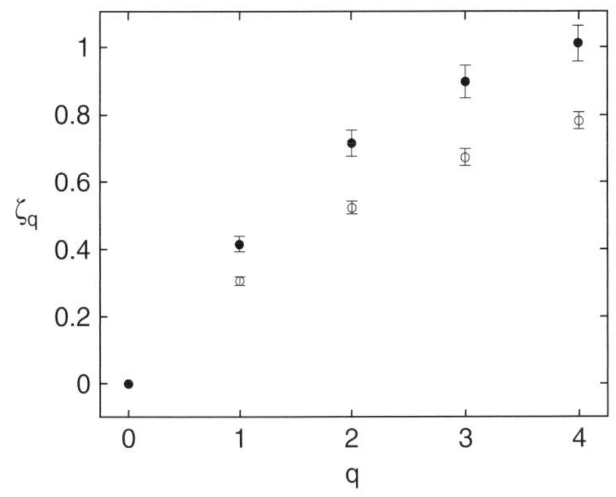

Figure 4.5. Structure function exponents using ESS for the time series of the strength of the magnetic field collected for extended periods during solar minimum (low solar activity) by the WIND spacecraft (circles) and solar maximum (high solar activity) by the ACE spacecraft (filled circles). (From Bruno et al. 2007.)

are expressed in terms of the correlation length that diverges at the critical point (Dubrulle 1994). This procedure suggested by Benzi and colleagues is called "extended similarity scaling" (ESS).

Calculations for structure function exponents have been carried out for magnetic and velocity fluctuations in the solar wind, for example (Bruno et al. 2007; Burlaga 1991a, b, c; Carbone 1994; Hnat et al. 2002; Horbury 1995a, b; 1997; Marsch and Tu 1997; Pagel and Balogh 2001; Tu et al. 1996), and elsewhere in the space environment, usually by assuming the applicability of the Taylor hypothesis. We caution here again that except for most of the solar wind studies, the assumption of the Taylor hypothesis must be exercised with a critical eye. In Figure 4.5, the results for such calculations for the time series of the strength of the magnetic field collected for extended periods during solar minimum (low activity) and solar maximum (high activity) by two separate spacecraft WIND and ACE (Advanced Composite Explorer launched on August 25, 1997 to study particles in space ranging from the solar wind, interplanetary, interstellar, and beyond) are presented (Bruno et al. 2007). We note that these structure function exponents exhibit nonlinear, that is, multifractal, behavior as expected for intermittent turbulence.

Structure function calculations may be performed conveniently for a fluctuating series for positive values of q; but will generally diverge for $q < 0$. The structure function exponents are easily proven to satisfy the following inequalities for $q_1 \le q_2 \le q_3$ provided the structure functions are defined by Eq. (4.20), (Biskamp 2003):

$$(q_3 - q_1)\zeta_{q_2} \ge (q_3 - q_2)\zeta_{q_1} + (q_2 - q_1)\zeta_{q_3} \tag{4.21}$$

$$\zeta_{q_1} \le \zeta_{q_2} \tag{4.22}$$

If the absolute sign over the incremental change is removed from (4.20), then the inequalities hold only for even positive integers of q. And it is easily verified that the structure functions are related to the spatial correlation functions.

4.5.3 A Formal Structure Function Description of Multifractals

Parisi and Frisch (1985) pointed out that the basic difference of monofractals and multifractals arose from the distinction of global- and local-scale invariants. If the dynamical complexity exhibit monofractal behavior, one global singularity suffices to characterize its invariant behavior. However, for multifractals, the complexity behavior is characterized by a weaker local invariance property of a spectrum of singularities. Formally, we may consider a variable fractal dimension spectrum $\bar{D}(h)$ that is dependent on the continuum singularities h of the local invariants of the subsets of fluctuations. Thus, the structure function of the qth moment order may be expressed as:

$$S_q = \int_I dh' \rho(h') \Delta^{qh' - \bar{D}(h')} \tag{4.23}$$

where I denotes the full set of fluctuations, and $\rho(h)$ gives the weight of the singularity h.

(Up to this point, power-law invariants were discussed from the point of view of linear relations of log-log plots over some small range of scales. In this subsection and some of the discussions in the following text, a more stringent requirement of power-law singularities in the limit of vanishing scale size is demanded.)

In the limit of $\Delta \to 0$, the power law with the smallest exponent dominates. Assuming $\rho(h') \neq 0$ and $\bar{D}(h')$ are smooth, the minimum is located at $h' = h(q)$ with

$$\bar{D}'(h(q)) = q \quad \text{and} \quad \bar{D}''(h(q)) < 0 \tag{4.24}$$

Therefore, based on this method of steepest decent, the qth order structure function for very small Δ is approximately

$$S_q \sim \exp\{[qh(q) - \bar{D}(h(q))] \ln \Delta\} \tag{4.25}$$

This leads immediately to the identification of

$$\zeta_q = [qh(q) - \bar{D}(h(q))] \tag{4.26}$$

Differentiating (4.26) with respect to q and using the first expression of (4.24), we obtain

$$h(q) = d\zeta_q / dq \tag{4.27}$$

Thus, given $\zeta_q(q)$, we may determine $h(q)$ and then, $\bar{D}(h)$. We note that the pairs (ζ, q) and (D, h) are Legendre transforms of each other. This formal identification of how the structure function exponent is related to the multifractal spectrum $\bar{D}(h)$ is, however, dependent on the statistics of the complete set of fluctuations that encompass all the singularities of the dynamical complexity; thus, the results may not be very accurate when performed numerically in practical applications. A more succinct procedure would be to identify the multifractal spectrum in a continuum fashion based on the statistics of the local invariance properties in terms of the subsets of fluctuations that define the local singularities. We shall discuss such a procedure, called ROMA (rank-ordered multifractal analysis), in Chapter 6.

4.5.4 Log-Poisson Model of Intermittency for the Structure Function Exponents

There exists in the literature a suite of phenomenological models designed to describe the intermittency characteristics of the nonlinear q-dependence of the structure function scaling exponent $\zeta(q)$ such as the log normal, the β, the random β models. These models have been discussed and explained in detail for hydrodynamic turbulence by Frisch (1995) and for MHD turbulence by Biskamp (2003).

Among these, one of the most successful data-fitting procedures is based on a model suggested by She and Lévêque (1994), which we will consider. It was suggested that such a model may be deduced by assuming the scale-dependent fluctuations obeying log-Poisson statistics (Dubrulle 1994; She and Waymire 1995).

Briefly, let us consider the probability distribution function $P(X_\ell)$ of the generic incremental scale (ℓ) dependent dimensionless fluctuations $X_\ell = \overline{X}_\ell / \overline{X}_\ell^\infty$ (where $\overline{X}_\ell^\infty = \lim_{p\to\infty} \langle \overline{X}_\ell^{p+1} \rangle / \langle \overline{X}_\ell^{p} \rangle \sim \ell^{-x}$ is a quantity tracing the tail of the distribution of a special class of raw random variables $\overline{X}_\ell$ in terms of the moments with a scaling exponent "$-x$") and define an associated random variable $Y = \log X_\ell / \log \beta$, which is assumed to satisfy the generalized Poisson distribution as follows:

$$F_{\lambda(\ell)}(Y) = \int e^{-\lambda} \frac{\lambda^z}{z!} G(Y-Z)dZ \tag{4.28}$$

where $F_{\lambda(\ell)}(Y)dY = P_{\lambda(\ell)}(X_\ell)dX_\ell$ and G is an arbitrary PDF.

Calculating the qth moment of the fluctuations, we obtain

$$\langle X_\ell^q \rangle = B_q \left[\langle X_\ell \rangle / B_1 \right]^{\frac{1-\beta^q}{1-\beta}} \tag{4.29}$$

where $B_q = \int dY' \beta^{qY'} G(Y')$, and for the scaling exponents:

$$\zeta_q = -qx + (\zeta_1 + x)(1-\beta^q)/(1-\beta) \tag{4.30}$$

Therefore, ζ_q depends on three fitting parameters, (ζ_1, x, β). Under such a situation, the nonlinear dependence of ζ_q on q is mainly in the regime of low values of q (for $\beta < 1$), which may be estimated by the dominant fractal structures of the fluctuations and other dimensional constraints.

One expects that fitting the experimental data to such an expression (4.30) can be quite robust because, in addition to the arguments given, the linear asymptotic behavior of ζ_q for large q is basically assured by the limitation of the availability of data residing in the extended wings of the probability distributions.

Dubrulle (1994) demonstrated this by identifying X_ℓ as the nondimensionalized expression of the so-called energy dissipation averaged over a ball of scale size ℓ and then relating it back to the longitudinal incremental velocity fluctuations for hydrodynamic turbulence; thus yielding an expression originally suggested by She and Lévêque (1994). This ansatz, also verified by She and Waymire (1995) using the log Poisson statistics, left essentially no apparent adjustable parameters in the scaling expression and confronted experimental data surprisingly well. This idea was followed by Politano and Pouquet (1995) for the structure function exponent spectrum for MHD intermittent fluctuations, also quite satisfactorily.

5

Partition Functions, Generalized Dimensions, and Singularity Spectra

5.1 Partition Function

In the previous chapter, probability distributions were analyzed as functions of the sizes of the incremental changes and scales. In principle, we may also partition the fluctuations into physical boxes (in time or space), or other abstract domains at different scales and study their probability distributions. The "multifractal" nature of the intermittent fluctuations may then be extracted from such distributions by searching for the "dominant singular behavior" of different moment orders at small scales analogous to that of the structure function exponents.

Based on the techniques suggested by Parisi and Frisch (1985) for structure functions as discussed in Section 4.2 and the generalizations suggested by Halsey and colleagues (1986), such ideas were first applied to the solar wind turbulence by Burlaga (1991a, b, c; 1993) and later for the auroral electrojet (AE) index by Consolini and colleagues (1996), for the high latitude geomagnetic fluctuations by Vörös (2000) and Vörös and colleagues (2003), for the plasma sheet by Lui (2001) and Weygand and colleagues (2005), to the magnetospheric cusp region by Yordanova and colleagues (2004), and to the 2D magnetohydrodynamic (MHD) simulations and the auroral electric field fluctuations by Chang and colleagues (2006).

Consider, for example, a generic observational time series $X(t)$ and define an "incremental measure" (Chang 2009):

$$\delta\mu_j = \left|X(t_j + \delta) - X(t_j)\right| \tag{5.1}$$

where δ is the sampling interval. (Although much of the ensuing discussion will be couched mainly along one-dimensional arguments, it will be straightforward to generalize the concepts and analyses to higher physical dimensions.) We now subdivide the total time interval T into $M = T / \tau$ segments with $\tau = k\delta$ and calculate the normalized scale-dependent "segmental measure"

$$\mu_i(\tau) = \sum_{j=(i-1)k+1}^{ik} \delta\mu_j / \mu \ with \ \mu = \sum_{j=1}^{N} \delta\mu_j \tag{5.2}$$

For convenience, we will generally choose M to be an integer.

The normalization of $\mu_i(\tau)$ is an attempt to represent the segmental measure as a measure of probability. This is the scale-dependent probability measure that we seek to define in terms of the segmental boxes. We shall now assume that each such measure varies with the scale τ in a singular manner as a power law for small values of τ and form the *qth* moment order of the coarse-grained probabilities $\mu_i(\tau)$, traditionally called the "partition function" (Halsey et al. 1986):

$$\Gamma(q, \tau) = \sum_{i=1}^{M} \mu_i^q(\tau) \tag{5.3}$$

where q may be any real number. We shall now search for the dominant singular behavior of $\Gamma(q, \tau)$ as characterized by a power law in τ with exponent $\gamma(q)$ for small τ similar to that for the structure function analysis. In general, for each moment order for intermittent fluctuations, the partition function exponent $\gamma(q)$ is a different number characterizing the particular fractal behavior of the subset of fluctuations, which dominates the (singular) scaling behavior of that particular moment order.

To demonstrate this idea more concretely, we follow Consolini and colleagues (1996) and consider the singularity measure multifractal analysis for a typical time series of the AE index. Following their calculations, the incremental measure is chosen as:

$$\delta\mu_j = \left|\phi_{AE}(t_j + \delta) - \phi_{AE}(t_j)\right|^2 \tag{5.4}$$

where $\phi_{AE}(t_j)$ is the measured AE index at time t_j and δ is the sampling interval. We note from Figure 5.1 that $\gamma(q)$ for this example is indeed a nonlinear function of q, and thus, a "multifractal." There are two apparent asymptotic regimes of q for sufficiently large values of $|q|$ for which $\gamma(q)$ varies with q nearly linearly and a nonlinear crossover region between the two asymptotes. We note that, unlike for the structure function case, it was possible to determine the partition function exponents for both positive and negative values of q without encountering undesirable divergences.

One may apply the same approach to the 2D MHD simulation data discussed in Chapter 4. The result is shown in Figure 5.2. Again, we note that there are apparently two linear asymptotes and a crossover regime. As mentioned previously, similar results have been found in various other space applications.

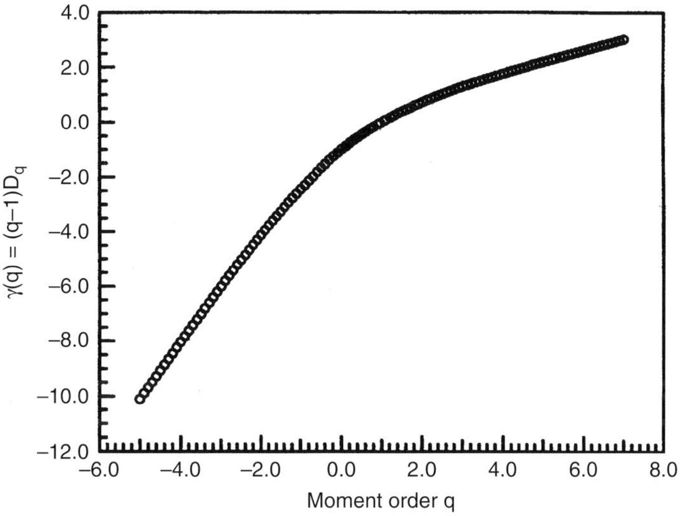

Figure 5.1. Partition function exponent $\gamma(q)$ for a time series of the AE index. (From Consolini et al. 1996.)

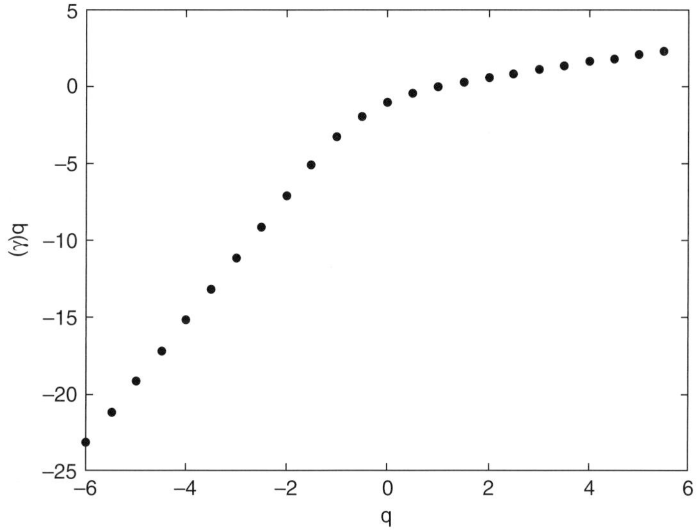

Figure 5.2. Partition function exponent $\gamma(q)$ versus q for the magnetic energy density fluctuations at small scales obtained from a snapshot in time of a 2D MHD homogeneous turbulence simulation on a $(512)^2$ periodic lattice.

A word of caution is in order here with regard to the generic feature of the straight line asymptotic regimes for large $|q|$. They appear mainly due to the lack of available data in the range of the simulated or observed results, similar to the reason for the asymptotic behavior for large orders of the structure function exponents of Chapter 4.

5.2 Generalized Dimensions

There exist various other approaches in the literature in expressing quantitatively the fractal characteristics of intermittent fluctuations. For example, in 1983, Henschel and Procaccia suggested an interesting concept of the so-called generalized dimensions for fractals. Their definition of the generalized dimension D_q may be related to the partition function $\Gamma(q)$ described in Section 5.1 as follows:

$$D_q = (q-1)^{-1} \lim_{\tau \to 0} \log \Gamma(q) / \log \tau = \gamma(q)/(q-1). \tag{5.5}$$

It is called the generalized dimension because for $q = 0$, D_0 is essentially the fractal (or box-counting) dimension of Mandelbrot (1977). Also, it may be shown that, in the limit of $q \to 1$,

$$D_1 = \lim_{\tau \to 0} \left(\sum_i \mu_i \log \mu_i \right) / \log \tau \tag{5.6}$$

where the term involving the summation is essentially the information or Shannon entropy as discussed in Section 4.1. And, for $q \geq 2$, D_q are the qth correlation dimensions. "D_q" has the property that it decreases continuously with the increase of the value of the index q.

The calculated generalized dimension D_q as a function of moment order q for the AE index discussed previously is shown in Figure 5.3. As demonstrated by Consolini and colleagues (1996) that such intermittency might be modeled by the so-called p-model as advanced by Meneveau and Sreenivasan (1987) for fully developed hydrodynamic turbulence. It is a special case of the "two-scale cascade model" as described by Halsey and colleagues (1986).

Briefly, the "two-scale cascade model" is constructed by hierarchically subdividing the one-dimensional unit interval [0, 1] with assigned probability measures. Initially, the interval is replaced (rescaled) by two intervals, l_1 and l_2, and assigned the measures p_1 and p_2, respectively, with

$$l_1 + l_2 \leq 1 \quad \text{and} \quad p_1 + p_2 \leq 1 \tag{5.7}$$

where all parameters are real. The new intervals are then rescaled repeatedly and self-similarly. Such set of numbers is classically called a Cantor set (Ott 1993). For the "p-model," $l_1 = l_2 = 1/2$ and $p_1 + p_2 = 1$. The generalized dimension of the p-model may then be calculated explicitly:

$$D_q = \log_2 [p^q + (1-p)^q]^{1/(1-q)} \tag{5.8}$$

As shown in Figure 5.3, with the choice of $p \equiv p_1 = 0.746 \pm 0.002$, the generalized dimension D_q as a function of moment order q for the AE index may be adequately represented by such a model.

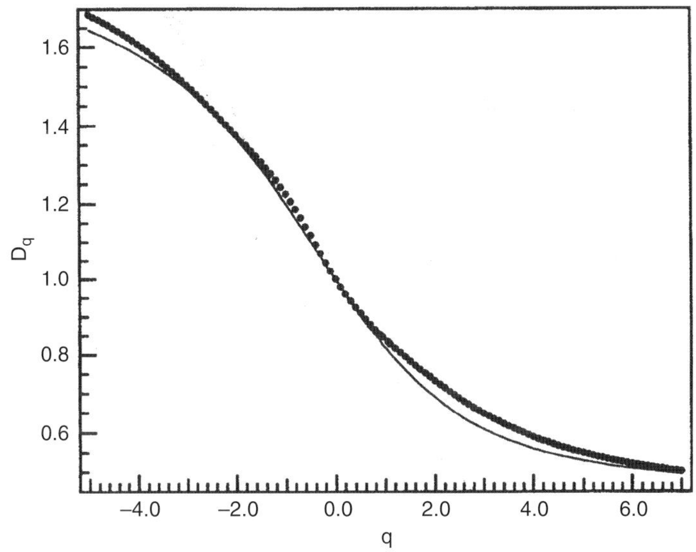

Figure 5.3. Generalized dimension $D(q)$ for the same time series of the AE index of Figure 5.1. Solid line is the best nonlinear fit by the P-model. (From Consolini et al. 1996.)

This one-parameter representation of the generalized dimension for the multifractal phenomenon of the AE index reinforces the idea stated earlier that the multifractal phenomenon can be thought of as the crossover phenomenon among invariant regions. Thus the concept of generalized dimensions is just another way of expressing the fractal dimensions for intermittent fluctuations.

It is common to define the intermittency coefficient as:

$$\mu_{\text{int}} = -2\left(dD_q / dq\right)_{q=0} \tag{5.9}$$

For $p = 0.746 \pm 0.002$, the intermittency coefficient is 0.400 ± 0.002. In reality, such a definition of intermittency will carry with it some physical meaning of intermittency representing the crossover behavior between two invariant properties such as those characterized by the p-model, but not much more than that.

5.3 The Singularity Spectrum f(α)

Another way to gauge the multifractal characteristics of intermittency is the singularity spectrum method (Halsey et al. 1986) which is very much related to the original idea of Parisi and Frisch (1985) expressed in terms of the structure function concept as discussed in Section 4.2. If we associate with each segmental measure μ_i of Section 5.1 a singularity index α_i such that $\mu_i = \tau^{\alpha_i}$ where α_i is within some range between α and $\alpha + \Delta\alpha$ and pass on to the continuum limit by writing the number of differential measures whose singularity index is between α and $\alpha + d\alpha$ as $\rho(\alpha)\tau^{-f(\alpha)}d\alpha$, then the partition function as defined in Eq. (5.3) in the continuum limit is:

$$\Gamma(q,\tau) = \int d\alpha' \rho(\alpha')\tau^{-f(\alpha')+q\alpha'}$$

$$= \int d\alpha' \rho(\alpha') \ell n(1/\tau) \exp[f(\alpha') - q\alpha'] \quad (5.10)$$

Because we are interested in the phenomenon for very small τ, the integral is dominated by the value of α' which makes the expression $f(\alpha') - q\alpha'$ the largest. Assuming $\rho(\alpha') \neq 0$ and $f(\alpha')$ smooth, the maximum is located at $\alpha' = \alpha(q)$ with

$$f'(\alpha(q)) = q \text{ and } f''(\alpha(q)) < 0 \quad (5.11)$$

Therefore the partition function for very small τ, using the method of steepest decent, is approximately

$$\Gamma(q,\tau) \sim \exp\{[q\alpha(q) - f(\alpha(q))]\ln \tau\} \quad (5.12)$$

This leads immediately to the identification of

$$\gamma(q) = D_q(1-q) = [q\alpha(q) - f(\alpha(q))] \quad (5.13)$$

Differentiating (5.13) with respect to q and using the first expression of (5.11), we obtain

$$\alpha(q) = \gamma'(q) \equiv d\gamma/dq \quad (5.14)$$

Thus, given $\gamma(q)$, we may determine $\alpha(q)$ and then $f(\alpha)$. Such a singularity spectrum for the AE index considered by Consolini and colleagues (1996) is given in Figure 5.4. Because $\alpha(q)$ is intimately related to $\gamma(q)$ (and therefore also to the partition function $\Gamma(q)$), $f(\alpha)$ sorts out the spectrum of singularities of the intermittent fluctuations, that may be mapped to the multifractal dimensions D_q as defined by (5.5).

In the preceding, we have introduced a sequence of algebraic manipulations mainly for the purpose of connecting three different popular approaches of identifying the singularity nature of the multifractal characteristics of a time (or spatial) series based on the partition function formulation. We must emphasize that these are merely different approaches of viewing fracture characteristics of intermittent fluctuations with no additional physics involved, except perhaps with the following useful thermodynamic analogy. That is, the relationship between $(\gamma(q), q)$ – or equivalently (D_q, q) – and $(f(\alpha), \alpha)$ is formally a Legendre transform. Thus one may easily associate the relationships among such measures with the standard

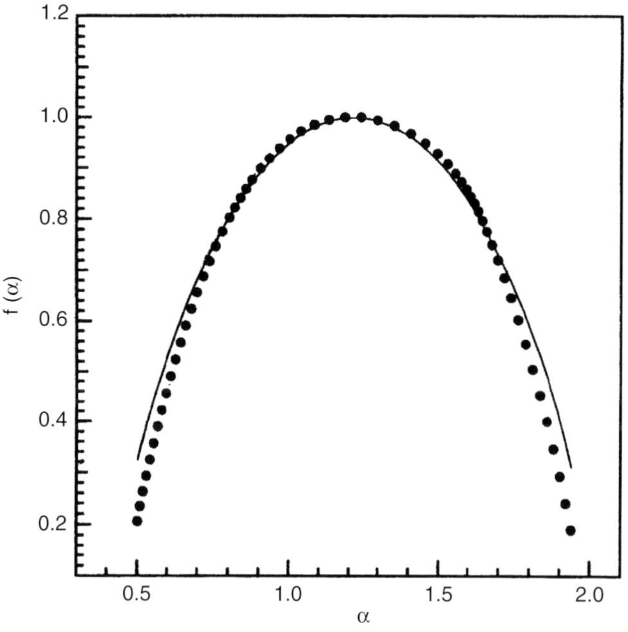

Figure 5.4. Singularity spectrum $f(\alpha)$ obtained from the Legendre transform of the partition function $\gamma(q)$ of Figure 5.3 for the time series of the AE index. Solid line is the best nonlinear fit by the P-model. (From Consolini et al. 1996.)

notion of Legendre transforms of classical thermodynamics (Ott 1993). Such a connection may become useful in identifying analogies between singularity measures of multifractals with multicriticality measures in thermodynamics.

It is interesting to note that in multifractal analyses of space plasma turbulence in regions where intermittency are expected, similar functional dependence of $\gamma(q)$ and $D(q)$ on q and $f(\alpha)$ as mentioned is also generally obtained (Burlaga 1991a, b, c; Carbone and Bruno 1996; Chang et al. 2006; Marsch et al. 1996; Weygand et al. 2005; Yordanova et al. 2004).

In terms of the specific p-model which Consolini and colleagues (1996) considered in Section 5.2, the singular spectrum $f(\alpha)$ fits nicely with the AE index result that they considered, Figure 5.4.

Degree of Multifractality and Measure of Asymmetry

Recently, based on the more general two-scale cascade model with $p_1 = 1 - p_2 = p$ and $l_1 = l_2 = \lambda \leq 1/2$, Macek (2007) and Macek and Wawrzaszek (2009) introduced the definitions of degree of multifractality $\Delta = \alpha_{\max} - \alpha_{\min}$ and measure of asymmetry $A = (\alpha_0 - \alpha_{\min})/(\alpha_{\max} - \alpha_0)$ where α_0 is the value of α for which $f(\alpha)$ is at its maximum and $\alpha_{\min}, \alpha_{\max}$ are the minimum and maximum values of α to characterize the basic aspects of the corresponding singularity spectrum of the model. By fitting the cascade model to actual observed data, these values (which are directly related to the fitting parameters of (p,λ)) provide some physical sense of the characteristics of the non-self-similarity of the analyzed intermittent fluctuations.

These authors and their collaborators (e.g., Lamy et al. 2010; Macek and Wawrzaszek 2011a, b; Macek et al. 2011) have applied this technique to the magnetic and velocity fluctuation data of the solar wind obtained by the ACE, Helios, Voyager 1 and 2, and Ulysses (Spacecraft that surveyed the solar wind in various regions of the heliosphere. The Voyagers launched in August and September 1977 surveyed the heliosphere and visited the major planets and are now heading off to the interstellar space. Ulysses, launched on October 6, 1994, is designed to study the heliosphere as a function of solar latitude.). Figure 5.5 gives examples of such results.

The solar wind travels at supersonic speeds until it crosses a shock wave near the heliospheric boundary called the *termination shock*. Beyond the termination shock is the heliosheath. Recent analysis (Macek et al. 2012) based on fluctuations of the interplanetary magnetic field strength measured by Voyager 2 spacecraft in these regimes indicated that the multifractality of the fluctuations decreases with the increase of distance from the Sun and in the very distant heliosphere and the heliosheath is correlated with solar activity. In addition, the asymmetry of the multifractal spectrum changes at the termination shock. Also, the magnetic field fluctuations seemed to indicate that both the intermittency and asymmetry decrease significantly as the heliopause was approached (Macek et al. 2014).

5.4 Wavelet Transform Modulus Maxima

In Section 3.7, it was shown that the intermittency characteristics of a fluctuating series could be vividly demonstrated using colored power spectra of wavelet transforms. And Local Intermittency Measure (*LIM*) (Normalized Power) as well as Flatness were introduced to quantify the intermittency effects. Because the complete set of the coefficients of the wavelet transforms essentially represents the full fluctuating event, it is reasonable to inquire if it can be used to advantage to describe the full characteristics of intermittency of a fluctuating series. The answer is a definite *yes*. For example, wavelet transforms have been used to study the multifractal characteristics of time series related to intermittent turbulence in the plasma sheet by Vörös and colleagues (2004), and for the magnetospheric cusp by Yordonova and colleagues (2004) using the WTMM (Wavelet Transform Modulus Maxima) technique as described in the following text.

In Figures 1.2 and 3.10, we noted that the patterns of the wavelet transforms for intermittent fluctuations exhibit certain geometrical invariant properties. It has been demonstrated that the curves representing the local maxima of these patterns were sufficient for the determination of the invariant singular behavior of the fluctuations. In fact, the ensemble of such curves might be used to partition the fluctuating properties in analogy to the box-sampling technique described in the previous section.

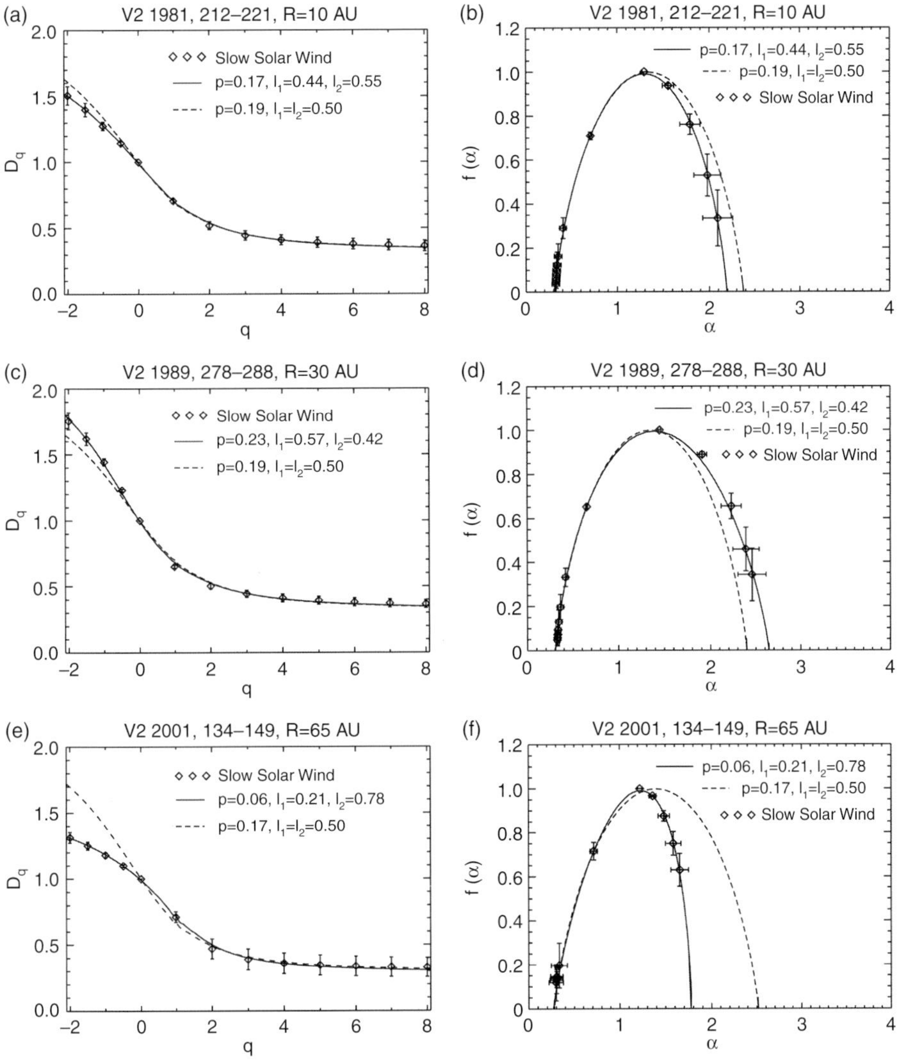

Figure 5.5. The generalized dimensions D_q (a, c, e) and singularity spectra $f(\alpha)$ (b, d, f) calculated for the one-scale p-model (dashed lines) and the generalized two-scale (continuous lines) models generated by the transfer rate of the energy flux based on the third moment of the longitudinal velocity fluctuations with parameters fitted to the multifractal measure using data measured by Voyager 2 during solar maximum (1981, 1989, 2001) at 10, 30, and 65 AU (diamonds) for the slow solar wind. (From Macek and Wawrzaszek 2011.)

These ideas were originally suggested by Mallet and Zhong (1992) and Mallet and Hwang (1992).

The central point of this method is based on the proof that the singular behavior for each localized wavelet pattern is contained within the line of local modulus maxima (maxima line) of the wavelet pattern. The singularity is characterized by

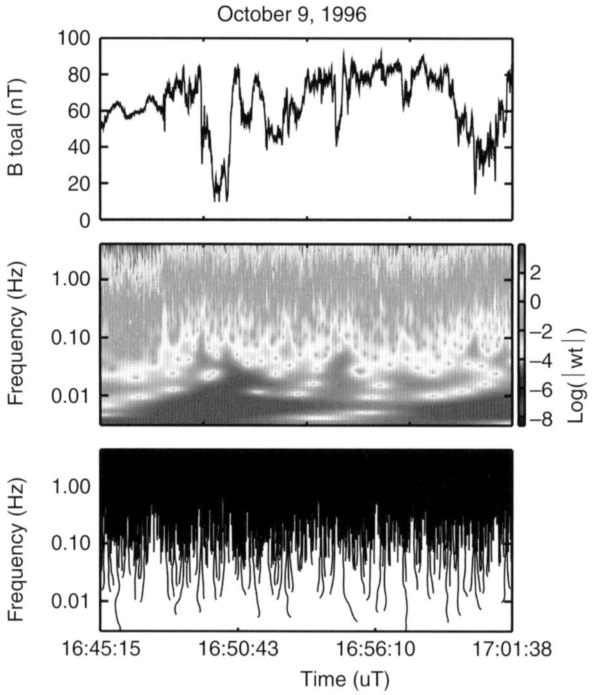

Figure 5.6. (Top panel): Segment of magnetic field data obtained by the Polar satellite on October 9, 1996 from 16:45:15 to 17:01:38 UT of the northern cusp region. (Middle panel): Its spectrogram. (Lower panel): Map showing the WTMM. (From Yordanova et al. 2004.) (For a color version of this figure, please see the color-plate section.)

the limit of the variation of the maxima at the vanishing scale size a of the transform. Thus, we can form a partition function based on such lines ℓ in analogy to the box-sampling technique of Eq. (5.3) as follows:

$$Z(q,a) = \sum_{\ell \in L(a)} \left(\sup_{a' \le a} \left| T_\psi[g](b_\ell(a'), a') \right| \right)^q \tag{5.15}$$

where $L(a)$ is the set of all the lines of modulus maxima at scale a, $T_\psi[g](b,a)$ is the wavelet transform, and b is the translation parameter. The supremum (sup) is the least upper bound or roughly speaking the maximum value of the quantity represented by the absolute sign.

It was shown by Muzy and colleagues (1994), Mallet and Hwang (1992), and Bacry and colleagues (1993) that each maxima line points toward a point $b_\ell(0)$ such that

$$T_\psi[g](b_\ell(a), a) \sim a^{\alpha(b_\ell(0))} \tag{5.16}$$

and that the number of maxima lines at scale a associated with the same exponent α scales like $a^{-f(\alpha)}$. Thus, again in analogy to the continuum formulation based on the partition in terms of boxes of Eq. (5.10), we obtain:

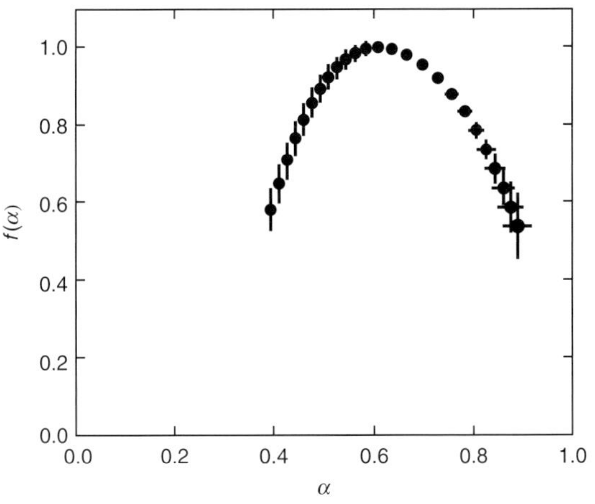

Figure 5.7. Singularity spectrum $f(\alpha)$ obtained with the WTMM method and Legendre transform for the cusp magnetic field fluctuations of Figure 5.6 (From Yordanova et al. 2004.)

$$Z(q,a) \sim \sum_{\alpha} a^{(q\alpha - f(\alpha))} \tag{5.17}$$

Thus, if $\gamma(q)$ is the scaling exponent characterizing the power-law behavior of the partition function, we obtain based on the argument that is equivalent to the method of steepest decent of Section 5.3, that

$$\gamma(q) = \inf_{\alpha}(q\alpha - f(\alpha)) \tag{5.18}$$

where infimum (inf) is the largest lower bound or roughly speaking the minimum of the quantity represented by the parenthesis.

Our analogy to the formalism presented in Section 5.3 is now complete and the singularity spectrum $f(\alpha)$ based on WTMM may be evaluated using the exponent $\gamma(q)$ based on the maxima line partition function and the Legendre transform.

Such a singularity spectrum for a particular set of Polar magnetic field fluctuation measurements in the magnetospheric cusp was obtained by Yordanova and colleagues (2004). The cusp is characterized by the magnetospheric region through which plasmas from the solar wind can have direct access to the upper atmosphere of the Earth. Plasmas in this region are highly turbulent. Figure 5.6 shows a segment of the data analyzed using the wavelet transform, the corresponding spectrum, and the map showing the locations of the modulus maxima. The mother wavelet employed in the analysis was the 4th order derivative of the Gaussian function. The calculated $f(\alpha)$ using WTMM and Legendre transform is shown in Figure 5.7, which has the characteristic shape of an inverted U with a single hump.

6

Rank Ordered Multifractal Analysis

6.1 Introduction

In Chapter 4, we considered a popular method of intermittency analysis based on the structure functions and discussed its connection with the concept of multifractals and local invariants using the Parisi-Frisch formalism. We also explained in detail the popular log-Poisson phenomenological description for the spectrum of the structure function exponents. These methods of analysis were based on the ensemble averages and moments of the full set of observed or simulated fluctuations even at times when the attention was focused on the singularities or scaling properties of the individual fractals.

Because most of the observed or simulated intermittent fluctuations are dominated by fluctuations with small amplitudes, the subdominant fractal characteristics of the minority fluctuations – generally of larger amplitudes – are easily masked by those characterized by the dominant population. It therefore appears reasonable to search for a procedure that explores the singular nature (and more generally the fractal, i.e., power-law scaling behavior) of the subdominant fluctuations by first appropriately isolating out the minority populations and then perform statistical investigations for each of the isolated populations and their scaling properties. We describe in this chapter a procedure ROMA (rank-ordered multifractal analysis) (Chang and Wu 2008) that is both physically explicable and quantitatively accurate in describing the spectrum of the multifractal characteristics of intermittent events in terms of the fractal natures of the individual groupings of the rank-ordered fluctuations based on the local invariants. The spectrum suggests a crossover phenomenon spanning from the near Gaussian characteristics of small amplitude fluctuations to the extreme intermittent state of large rare events. In the intermediate range of amplitudes, the fluctuations spread into a continuum of fractals. The generic nature of such rank-ordered multifractal spectrum leads it to a natural connection with the one-parameter scaling idea for monofractals.

To be more specific, let us consider a generic probability distribution function (PDF), $P(X,\delta)$, of fluctuations X for a range of scales δ, similar to those discussed in Chapter 4 and study how the preceding idea may be achieved with a rank-ordered method that subdivides the fluctuations into groupings based on their local invariants. We recall that such PDFs were generally non-Gaussian for small scales and became more and more heavy-tailed at smaller and smaller scales. In Chapter 4, we considered the attempt to collapse such unscaled PDFs according to the one-parameter scaling formula that had the following form:

$$P(X,\delta)\delta^s = P_s(X/\delta^s) \tag{6.1}$$

with a single scaling exponent s. The scaling formula may be interpreted from the concept of fractals. We begin such an interpretation by assuming that X and P vary with δ as power laws: $X/\delta^a = I$ and $P/\delta^b = J$, where (a, b) are the power-law (fractal) exponents and (I, J) are constants – that is, invariants with respect to the change of scale δ. If the functional form of $P(X,\delta)$ is also invariant as the scale changes, then it has been shown that a functional relation exists between the two invariants, (I, J) (Chang et al. 2004). Imposing the normalization condition for the PDFs, we obtain the one-parameter scaling form with $s = a = -b$, a result originally obtained in Chapter 4 using the formalism of generalized homogeneous functions (GHFs).

Such a phenomenon is self-similar and the PDFs collapse onto one scaled PDF $P_s(Y)$ where $Y = X/\delta^s$ is a global scale invariant. And, the structure function exponents for the PDFs can be shown to satisfy the monofractal property of $\zeta_q = sq$.

The scaling exponent s (popularly known as the Hurst exponent) may be interpreted as the single fractal measure that characterizes the fluctuations of all scales according to the scaling relation (6.1). If the PDFs are self-similar Gaussian distributions for all scales, representing random diffusion, the scaling exponent s may be shown to be 0.5. For other monofractal distributions, the scaling exponent can take on any positive real value.

For intermittent fluctuations, such one-parameter globally invariant scaling is not satisfied. And the structure function calculations based on the full set of simulated or observed fluctuations generally exhibit a nonlinear relation between the exponents and the moment order, indicating the phenomenon is multifractal in nature (such as that described by Parisi and Frisch, 1985). However, the physical interpretation of the multifractal nature is not easily deciphered by merely examining the curvatures of the deviations from linearity.

We shall therefore attempt to perform statistical analyses individually for subsets of the fluctuations that characterize the various fractal behaviors within the full multifractal set of intermittent fluctuations. Such grouping of fluctuations must depend somehow on the sizes of the fluctuations. However, the groupings cannot depend merely on the raw values of the sizes of the fluctuations because the ranges will be different for different scales. Thus we are led to proceed to rank-order the sizes of the fluctuations based on the local invariant $Y = X / \delta^s$ where s is the scaling exponent for each (local) grouping.

Consider a differential range of dY in the vicinity of some scaled size $Y = X / \delta^s$. We expect the fluctuations whose sizes fall within this differential range to exhibit monofractal behavior characterized by the local scaling exponent s such that the differential structure function dS_q will vary with the scale as δ^{sq} according to:

$$dS_q \triangleq X^q P(X,\delta)dX = \delta^{sq} Y^q P_s(Y)dY \tag{6.2}$$

Given an ensemble of PDFs $P(X,\delta)$, the corresponding multifractal spectrum $s(Y)$ may be obtained approximately (if the ansatz is valid) by integrating the functional differential expression (6.2) over small contiguous ranges of ΔY with the assumption that within each incremental range the scaling exponent s is essentially a constant. Thus, for a range of ΔY within (Y_1, Y_2), we form a range-limited structure function as follows:

$$\Delta S_q(X,\delta) = \int_{a_1}^{a_2} X^q P(X,\delta)dX \simeq \delta^{sq} \int_{Y_1}^{Y_2} Y^q P_s(Y)dY \tag{6.3}$$

where $a_1 = Y_1\delta^s$ and $a_2 = Y_2\delta^s$. We may then search for the value of s such that the scaling property of the range-limited structure function that varies with s is $\Delta S_q(s) \sim \delta^{sq}$. If such a value of s exists, then we have found one region of the multifractal spectrum of the fluctuations such that the PDFs in the range of ΔY collapses onto one scaled PDF. Performing this procedure for all contiguous ranges of ΔY will produce the approximate rank-ordered multifractal spectrum $s(Y)$ that we are looking for. The determined value of s for each grouping should be unaffected by the statistics of other subsets of fluctuations that are not within the chosen range ΔY and therefore should be quantitatively quite accurate. If this spectrum exists, the PDFs for all time lags collapse onto one master multifractal scaled PDF, $P_s(Y)$. The spectrum will be implicit because Y is defined as a function of s (the local Hurst exponent).

The preceding discussion yields a ROMA spectrum that is stepwise discontinuous. One might wish to improve the calculated result by progressively decreasing the size of ΔY. However, this procedure may be limited by the statistics within each ΔY due to the limitation of available data. Therefore, a refined method of obtaining a continuum of the spectrum $s(Y)$ and the associated scaled PDF $P_s(Y)$ was suggested by Wu and Chang (2011) as described briefly in the following text.

A Refined Procedure

We can write the ROMA scaling relations as:

$$P(X,\delta)\delta^{s(Y)} = P_s(Y) \text{ with } Y = X / \delta^{s(Y)} \tag{6.4}$$

Thus, for a given value of s, we may plot $P(X,\delta)\delta^{s(Y)}$ against $Y = X / \delta^{s(Y)}$ for the various scales δ of the PDFs. If the curves intersect at some point Y_1 then ROMA is satisfied at $s = s(Y_1)$. However, if only curves within some range of scales δ intersect, then ROMA is satisfied at $s = s(Y_1)$ only for that range of scales. (This, in fact, sets the stage for scenarios where there are multiple ROMA scaling ranges as we shall discover.)

Continue this procedure for the full range of values of s then leads us to a continuum ROMA $s(Y)$ as well as the corresponding scaled PDF $P_s(Y)$. We may then fine-tune the result numerically by calculating the PDFs from $s(Y)$ and $P_s(Y)$ and compare them with the original observed or numerically simulated PDFs.

Such a procedure assures that the obtained results are unique. In other words, the continua of $s(Y)$ and $P_s(Y)$ represents the full ensemble of the raw PDFs of the observational or simulated results and vice versa.

ROMA has several advantages over the results obtained using the conventional structure function calculations. First, the utility of the spectrum is to fully collapse the unscaled PDFs. Second, the physical interpretation is clear. It indicates how space filling (in terms of the value of s) the scaled fluctuations are once the value of Y is given. Third, the determination of the values of the fractal nature of the grouped fluctuations is not affected by the statistics of other fluctuations that do not exhibit the same fractal characteristics. Fourth, it provides a natural connection between the one-parameter scaling idea (6.1) for monofractals and the multifractal behavior of intermittency.

Finally, the ROMA procedure emphasizes the fact that fractals are defined in terms of strict power-law invariants and not limited only to ranges of small scales. Some of the earlier discussions related to the analyses of multifractals require the fractals defined in terms of power-law singularities in the limit of small scales. Those results are mathematically beautiful but not necessarily very practical, particularly when local invariant concepts are required.

In the following sections, we shall discuss several numerical and observational examples of ROMA and its generalizations to multiple rank-ordered parameters with space applications.

6.2 Examples of ROMA Based on Numerical Simulations

6.2.1 2D MHD Simulation

Our first example of ROMA is based on the results of a 2D magnetohydrodynamic (MHD) simulation discussed by Chang and colleagues (2004) and Chang and Wu (2008) similar to those considered in the previous chapters. Briefly, in the simulation, ideal compressible MHD equations expressed in conservative forms are numerically solved with 1024×1024 grid points in a doubly periodic (x, z) domain of length 2π in both directions using the WENO code (Jiang and Wu 1999) so that the total mass, energy, magnetic flux, and momenta are conserved. The initial condition consists of a random magnetic field and velocity with a total pressure representing a high beta plasma. Periodic boundary conditions are applied. After sufficient elapsed time, the system evolves into a set of randomly interacting multiscale coherent structures exhibiting classical aspects of intermittent fluctuations such as that displayed in Figure 3.1.

Spatial values of the square of the strength of the magnetic field B^2 are collected over the entire (x, z) plane (with approximately 10^6 data points) at $t = 300$ and analyzed. For reference, the sound wave and Alfvén wave traveling times through a distance 2π are about 4.4 and 60, respectively. The PDFs $P(|\delta B^2|, \delta)$ are non-Gaussian and become more and more heavy-tailed at smaller and smaller scales. As discussed previously, an attempt to collapse the unscaled PDFs according to the monofractal scaling formula that is analogous to Eq. (6.1) indicated approximate global scaling with an estimated scaling exponent $s \simeq 0.34$, Figure 4.4. And structure function calculations exhibited a nonlinear relation between the exponents and the moment order.

Because for this example the PDFs exhibited approximate monofractal scaling, such structure function exponent calculations – though indicating multifractality – are strongly masked by the statistics of the population of the smaller fluctuations. Thus, this example represents an ideal candidate to test the utility of ROMA.

We now follow the procedure suggested in Section 6.1 and search for the ROMA spectrum $s(Y)$ and the scaled PDF $P_s(Y)$ where $Y = |\delta B^2| / \delta^s$. Such a stepwise approximate spectrum was found by Chang and Wu (2008). In Figure 6.1 a, b, we give the continuous functions of $s(Y)$ and $P_s(Y)$ using the refined method as described at the end of Section 6.1 (Wu and Chang 2011). We note that the continuous ROMA spectrum (Figure 6.1 a) has values ranging approximately from 0.50 to 0.22. Thus, at $Y = 0$, the scaling exponent is equal to the Gaussian value of 0.5 for random diffusion. As the value of the local invariant Y increases, the value of the scaling exponent s decreases continuously indicating the fluctuations are becoming less and less space-filling, that is, the diffusion phenomenon related to the fluctuations is becoming more and more antipersistent. Figures 6.1 c, d demonstrate the truly remarkable result that the full ensemble of the PDFs of the simulated 2D MHD intermittent turbulent fluctuations are indeed faithfully represented by the two continuous functions of the local invariants, $S(Y)$ and $P_s(Y)$.

6.2.2 Fluid Turbulence

In this second example, we apply ROMA to the inertial range of isotropic fluid turbulence (Wu and Chang 2011). The analysis is based on the vast simulation data made available by the Johns Hopkins University (Li et al. 2008; Perlman et al. 2007).

The data is obtained from a large-scale numerical simulation of driven isotropic turbulence on a $(1024)^3$ periodic grid in a physical domain of $(2\pi)^3$ using the

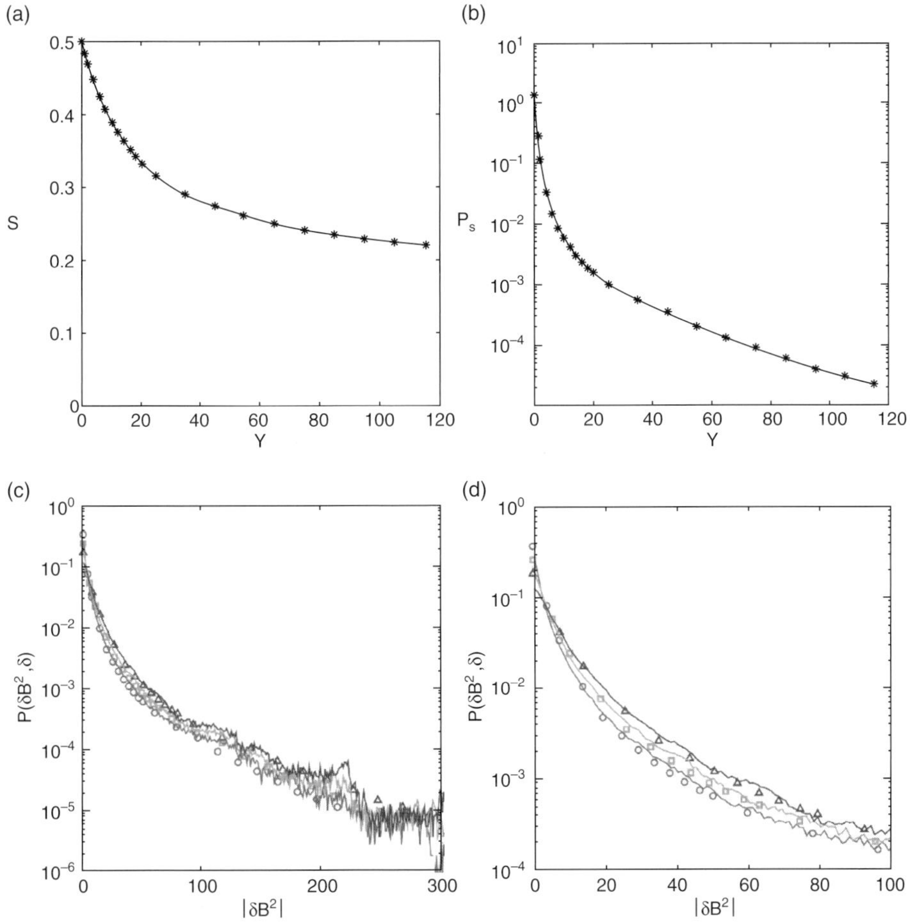

Figure 6.1. Fluctuations of B^2 based on a 2D homogeneous MHD simulation. (a) ROMA spectrum, $s(Y)$. (b) Scaled PDF, $P_s(Y)$. (c) Plots of computed PDFs from the ROMA relations (markers) and from the MHD simulation (solid curves). $\delta = 32\Delta$ (circles), 48Δ (squares): 96Δ (triangles). Δ (grid spacing) $= 2\pi / 1024$. $|\delta B^2|$ in units of $|\delta B^2_{\max}| / 800$. (d) Enlarged view of (c) for $|\delta B^2| \le 100$. (From Wu and Chang 2011.) (For a color version of this figure, please see the color-plate section.)

pseudospectral parallel code. Energy is injected by keeping constant the total energy in modes such that their wave-number magnitude is less or equal to two. After the simulation has reached a statistical stationary state, 1024 frames at every ten time steps of data, which includes the three components of the velocity vector and pressure, are generated. The duration of the stored data was about one large-eddy turnover time of 2.024. The radial energy spectrum averaged over this duration indicated the existence of an inertial range of wave numbers approximately between eight to sixty, corresponding to a spatial range from 17Δ to 128Δ where $\Delta = 2\pi / 1024$ is the grid spacing. For the ROMA analysis, nineteen arbitrarily chosen z-planes of data for wave numbers within the inertial range at different times were studied. The fluctuations of the longitudinal velocity $\delta v_{\|} = (\mathbf{v}(\mathbf{r} + \delta\mathbf{i}) - \mathbf{v}(\mathbf{r}))\cdot\mathbf{i}$, where δ is the spatial

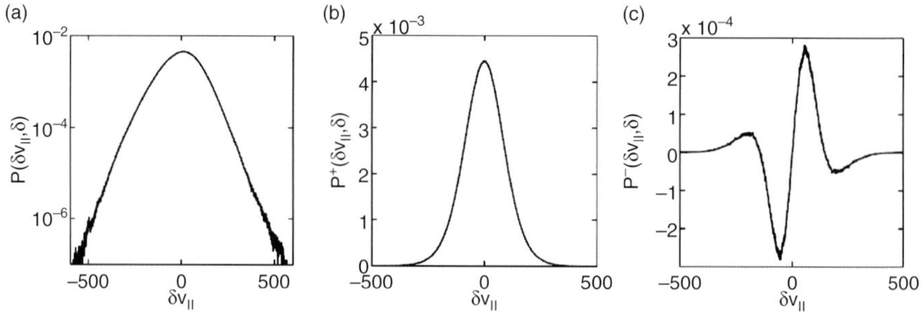

Figure 6.2. Longitudinal velocity fluctuations for fully developed isotropic fluid turbulence in the inertial range: (a) PDF of $\delta v_{\|}$ at $\delta = 64\Delta$ in units of bin size. (b) and (c) are the symmetrical and antisymmetric plots (P^{+}, P^{-}) of the PDF. (From Wu and Chang 2011.)

scale, are then analyzed. In the analysis **i** is either in the x- or y-axis and δ is in the range of $(16\Delta, 160\Delta)$.

Figure 6.2 gives the PDF results for $\delta v_{\|}$ at scale $\delta = 64\Delta$. In computing the PDF, the range of $\delta v_{\|}$ is divided into 1,601 bins. For bin number i: $(i-(1/2))\Delta_v < \delta v_{\|} \le (i+(1/2))\Delta_v$. The bin size Δ_v is set as 8/1,601 because the maximum value of $|\delta v_{\|}|$ is slightly less than four. The PDF is asymmetrical in $\delta v_{\|}$ and thus may be decomposed into a symmetrical and an antisymmetric part. The reason for the asymmetry is due to the 3D nature of the fluctuations.

Figures 6.3 shows the results of the analyzed continuous ROMA spectrum $s(Y)$ and the scaled PDFs. Interestingly, in spite of the asymmetric property of the PDFs, the ROMA spectrum is symmetric as shown. It shows approximate monofractal behavior at $Y \le 25$ and decreases monotonically at larger Y, indicating the turbulence is antipersistent and fully developed.

The PDFs are mapped onto a scaled master curve $P_s(Y)$ which is asymmetric as shown in the figure. It is interesting to note that such a scaled PDF may be fitted quite accurately by a Castaing distribution (Eq. 4.7) modified to accommodate the asymmetric property:

$$P_s(Y) = \frac{A(a_s)}{2\pi\lambda}\int_0^{\infty} \exp\left[-\frac{(Y-Y_0)^2}{2\sigma^2}\left(1 + a_s \frac{(Y-Y_0)/\sigma}{1+(Y-Y_0)^2/\sigma^2}\right)\right]$$

$$\exp\left[-\frac{\ell n^2(\sigma/\sigma_0)}{2\lambda^2}\right]\frac{d\sigma}{\sigma^2} \tag{6.5}$$

with $\sigma_0 = 18.25,\ Y_0 = 2{,}55,\ \lambda = 0.425,\ a_s = 0.4.$

The comparisons of the PDFs from the simulation data and those obtained from ROMA are shown in Figure 6.4. The results indicate that in the inertial range

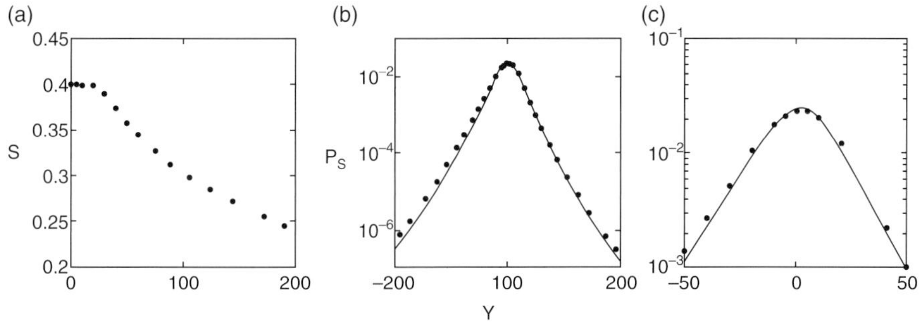

Figure 6.3. ROMA spectrum $s(Y)$ and scaled PDF $P_s(Y)$ for the fluctuations of longitudinal velocities of Figure 6.2. (a) $s(Y)$, which is symmetric with respect to $Y = 0$, shows approximate monofractal behavior at $Y \leq 25$ and decreases monotonically at larger Y. (b) Filled circles are calculated values for the $P_s(Y)$, which is asymmetric about $Y = 0$. Solid curves are fits with a modified asymmetric Castaing distribution with $\sigma_0 = 18.25$, $Y_0 = 2, 55$, $\lambda = 0.425$, $a_s = 0$. (c) is an enlarged view of (b) but with a smaller range of Y. (From Wu and Chang 2011.)

for isotropic fluid turbulence the PDFs exhibit multifractal scaling that can be described conveniently using ROMA. Conversely, this also means that the spectrum and scaled PDFs faithfully reproduce all the raw PDFs in the inertial range, a truly remarkable result.

6.3 Applications to Observations

6.3.1 Solar Wind Turbulence

As discussed in Section 4.1, solar wind turbulence is generally analyzed based on the Taylor hypothesis because the transit time of eddies (bundled fluctuations) is commonly much less than the characteristic evolution time. In addition, because the spacecraft speed is much smaller than the solar wind speed, the data gathered by instruments aboard the spacecraft essentially sample the spatial fluctuations in the direction of the solar wind, particularly at times when the solar wind turbulence is homogeneous at large scales and fully developed. Thus, for example, the time series of such measurements of the strength of the magnetic field $B(t_i)$, where $t_i = i\delta$ with $i = 0,1,2,....,N$ and δ is the sampling interval, may be interpreted based on the hypothesis as the spatial series of the ambient turbulence fluctuations with $x = v_S t$ (x: direction of the solar wind velocity). We can then construct the PDFs $P(\delta B^2, \tau)$ with the coarse-grained scale $\tau = k\delta$ (k: integer).

Non-Gaussian probability distribution functions from such statistics have been commonly observed in the solar wind for fluctuations related to the magnetic, velocity, density, and other field variables. Using a particular set of WIND data (Chang et al. 2010; Wu and Chang 2011), a ROMA spectrum $s(Y)$ for the PDFs $P(\delta B^2, \tau)$ was obtained, Figure 6.5 a, b. The spectrum had the signature of that of Figure 6.1 a.

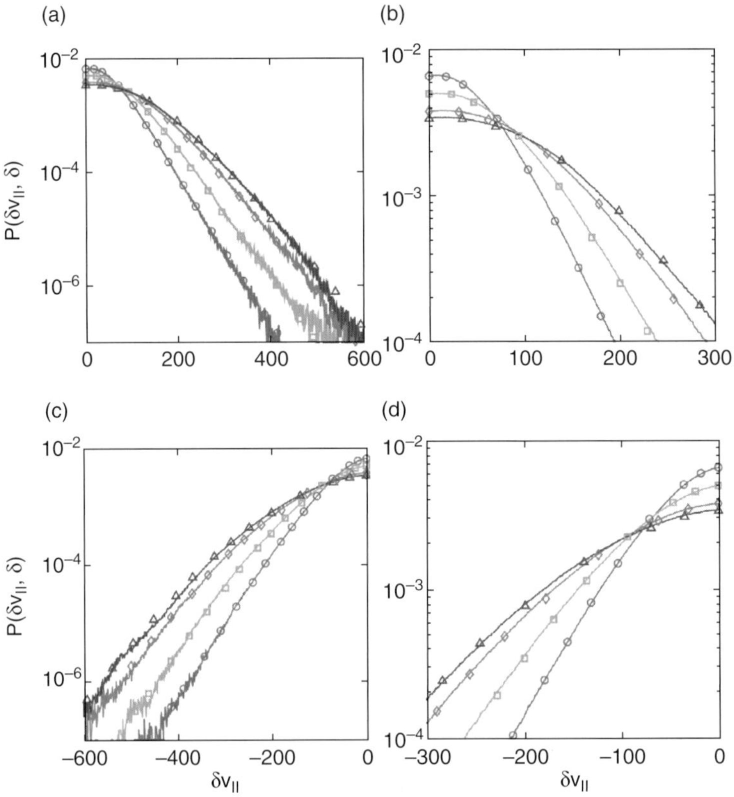

Figure 6.4. Plots of $P(\delta v_{\|},\delta)$ for fully developed isotropic fluid turbulence in the inertial range as discussed in Figures 6.2 and 6.3: solid curves are from the simulation data and markers are from ROMA scaling relations: circles for $\delta = 24\Delta$; squares for $\delta = 48\Delta$; diamonds for $\delta = 96\Delta$; triangles for $\delta = 128\Delta$. (a) and (b) are for positive $\delta v_{\|}$; (c) and (d) are for negative $\delta v_{\|}$. (b) and (d) are enlarged plots. (From Wu and Chang 2011.) (For a color version of this figure, please see the color-plate section.)

However, it was much flatter, that is, the range of values of the fractal index s (roughly between 0.44 and 0.38) for the full observed range of the scale invariant $Y \equiv \delta B^2 / \tau^s$ was quite small. In fact, it was found that the observed PDFs nearly collapse unto one single curve according to (6.1) by setting the fractal index s as a constant (~ 0.44), quite similar to the result obtained by Hnat and colleagues (2002), Figure 4.3.

We note from Figure 6.5c, the continuum ROMA spectrum $s(Y)$ and the scaled PDF $P_s(Y)$ faithfully reproduce the ensemble of the raw observational PDFs similar to the previous results obtained for the simulation examples. The fluctuations are antipersistent and the scaling exponent s decreases continuously with the increase of value of $|Y|$.

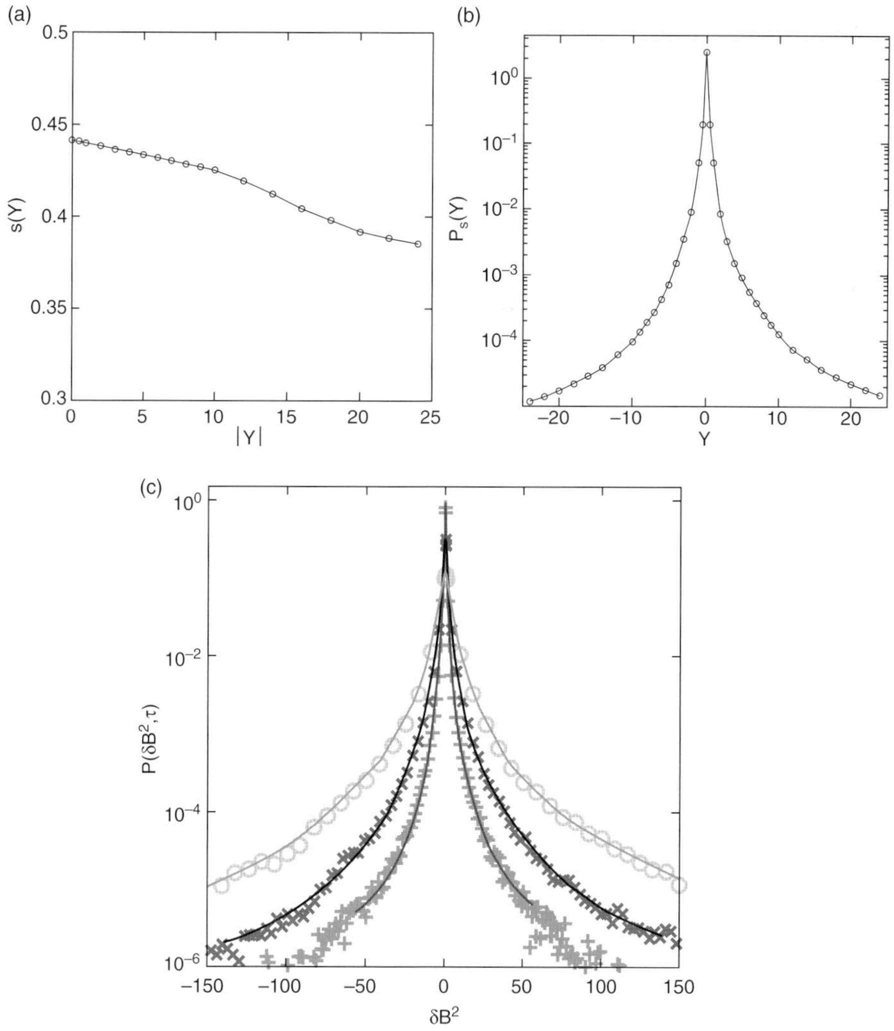

Figure 6.5. Fluctuations of B^2 based on the solar wind data. (a) and (b) ROMA spectrum $s(Y)$ and scaled PDF $P_s(Y)$. (c) Computed PDFs $P(\delta B^2,\tau)$ from ROMA relations (solid curves) and from solar wind observations (markers). δB^2 in units of $(nT)^2$. $\tau = 1000s$(o), $96s$(x). $9s$(+). (From Wu and Chang 2011.) (For a color version of this figure, please see the color-plate section.)

6.3.2 The Auroral Electrojet Index

In the previous chapter, it was shown that the intermittent fluctuations of the auroral electrojet (AE) index had multifractal characteristics. Recently, Consolini and De Michelis (2011) applied the ROMA technique to a long AE series covering a period of about 10.5 yrs from January 1, 1978 to June 30, 1988 for a total of nearly 6×10^6 data points. The choice of such a long time series was to

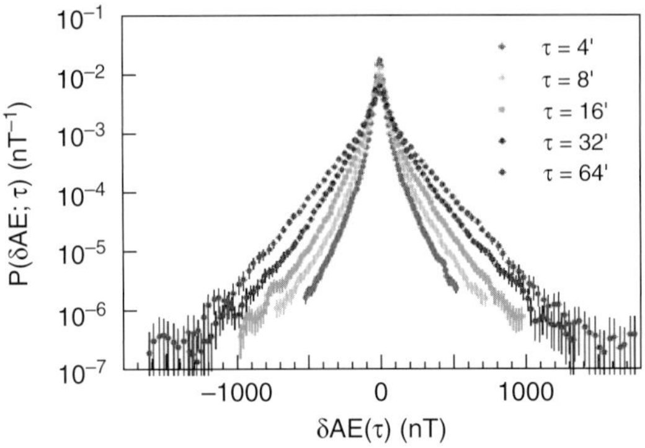

Figure 6.6. PDFs $P(\delta AE;\tau)$ for AE increments $\delta AE(\tau)$ at $\tau = 4,8,...,64$ minutes. (From Consolini and De Michelis 2011.) (For a color version of this figure, please see the color-plate section.)

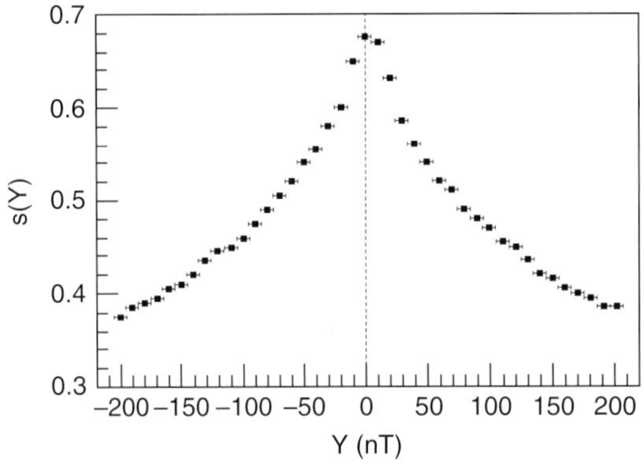

Figure 6.7. ROMA spectrum $s(Y)$ for the PDFs $P(\delta AE;\tau)$ of Figure 6.6. A slight asymmetry is noticeable in the range of $|Y| \leq 25$. Y is in units of nT. (From Consolini and De Michelis 2011.)

consider a period of the order of the solar cycle encompassing all the possible magnetospheric states.

In computing the PDFs, a nonparametric estimation based on the kernel technique was used (Kaiser and Schreiber 2002) in an attempt to reduce the errors that may arise from the discrete nature of the time series. And the kernel was chosen of the Gaussian type. Specifically,

$$P(\delta AE,\tau) = \frac{1}{N\varepsilon}\sum_{n=1}^{N} K(u) \tag{6.6}$$

where $u = (\delta AE - \delta AE(n))/\varepsilon$, $\delta AE(n)$ is the actual AE-increment time series, $\varepsilon = 0.2\sigma$ is the bin width, σ is the standard deviation of δAE, and $K(u) = (2\pi)^{-1/2}\exp(-u^2/2)$. Figure 6.6 displays the PDFs for temporal scales of $\tau = 4,8,...,64\,\text{min}$. We note their distinct non-Gaussian characteristics. The error bar estimates is performed assuming a Poisson distribution of points in $u \leq 1$.

The calculated discrete ROMA spectrum $s(Y)$ and scaled PDFs $P_s(Y)$ with $Y = \delta AE(\tau)/(\tau/\tau_0)^s$ with $\tau_0 = 60s$ are presented in Figures 6.7 and 6.8. Both are not symmetric with respect to $Y = 0$. The asymmetry is probably due to the difference between the increase (with positive AE increments) and decay (with negative

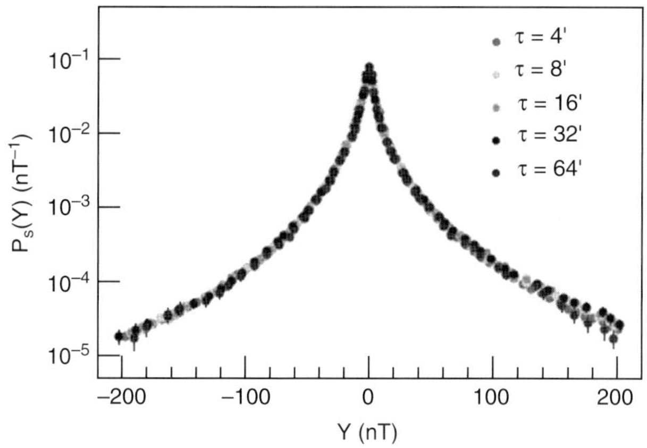

Figure 6.8. Scaled PDF $P_s(Y)$ for the probabilities of the AE increments of Figure 6.6 in units of $(nT)^{-1}$. Y is in units of nT. (From Consolini and De Michelis 2011.) (For a color version of this figure, please see the color-plate section.)

AE increments) of the electrojet current intensities. The fractal spectra $s(Y)$ for both the positive and negative Ys start out to be persistent for small $|Y|$ and decrease as $|Y|$ increases indicating the fluctuations became sparser and sparser as the magnitude of the local scaled invariant increases.

6.3.3 Turbulent Fluctuations in the Magnetospheric Cusp

As discussed in Section 5.4, the cusp is the magnetospheric region through which plasmas from the solar wind can have direct access to the upper ionized atmosphere of the Earth. Its magnetic field fluctuations have been shown to be turbulent, anisotropic (Yordanova et al. 2004; 2005), and sporadically intermixed with small scale Alfvénic vortexlike structures (Sundkvist et al. 2005). Detailed analyses (Echim et al. 2007) revealed the cusp as a transition region from a nonintermittent turbulent state inside the magnetosphere to an intermittent turbulent state in the magnetosheath (shocked solar wind) with statistical properties resembling the solar wind turbulence.

We study the ROMA characteristics of the magnetic energy density fluctuations in the region observed by Cluster (a constellation of four identical spacecraft in tetrahedral formation launched in 2000 by ESA with NASA participation traversing the vicinity and the interior of the Earth's magnetosphere) during a typical cusp passage, Figure 6.9. The time series data are interpreted as spatial fluctuations by assuming the validity of the Taylor hypothesis. An implicit partial removal of the dipole component of the magnetic field fluctuations is performed by computing first the differences $\delta B^2(t,\tau)$ from the raw data with τ being the time scale and then the mean value of the fluctuations at each scale subtracted yielding a new "ensemble" of fluctuations as follows (Echim et al. 2007),

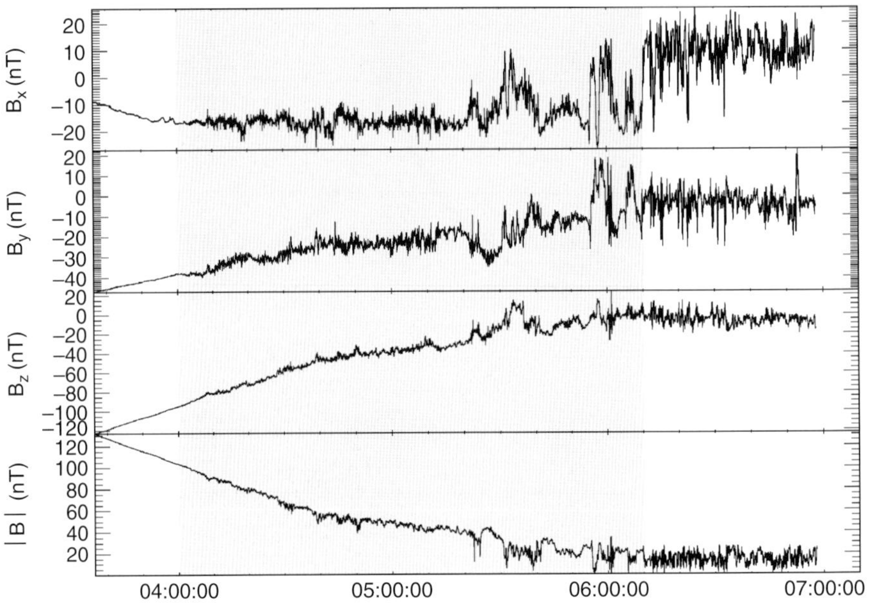

Figure 6.9. The components and intensity of the magnetic field measured by Cluster-1. The interval corresponding to the passage of the cusp region has been shaded in gray. (From Echim et al. 2007.)

$$\delta b^2 \equiv \frac{\delta B^2(t,\tau) - \langle \delta B^2(t,\tau) \rangle}{\sigma^2} \tag{6.7}$$

where the bracket indicates the ensemble average and σ is the variance.

Typical PDFs are computed for the quantity, δb^2, where differences δB^2 have been calculated by moving an overlapping window of width $\tau = 2^j \delta t$ over the entire time interval with $\delta t = 0.0015\,\text{sec}$ being the time resolution of the measurements and $j = 1,2,...,15$, Figure 6.10. The PDFs depart significantly from a Gaussian distribution with extended "wings" representing increased probabilities of large amplitude fluctuations at time scales smaller than $\tau_G = 61.47s$.

A ROMA using the approximate segmental integral technique was performed for the shaded data set in the cusp region (Chang et al. 2011; Lamy et al. 2008). We note from Figure 6.11 that for small values of Y, the fluctuations were persistent ($s > 0.5$), indicating the turbulence was unstable and probably not fully developed. For larger values of Y, the turbulence became well-developed and antipersistent ($s < 0.5$) with the fluctuations becoming sparser and sparser as the value of Y increased. The ROMA spectra for all four spacecraft were very similar; indicating that despite the fact the magnetic field fluctuations were anisotropic, the magnetic field energy density fluctuations in the cusp were essentially statistically isotropic over the distance covered by the cusp passage for spatial scales of the order of the separation distance between the Cluster spacecraft (~1000 km).

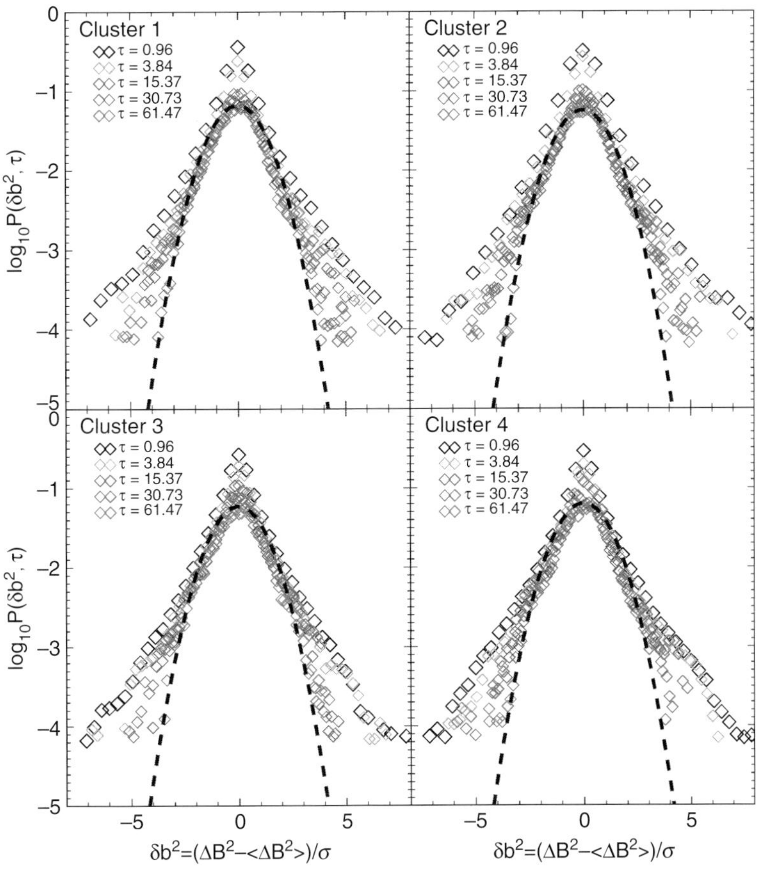

Figure 6.10. Typical PDFs of the magnetic energy density fluctuations measured by the four Cluster satellites for the cusp passage described in Figure 6.9; the PDFs have been scaled with respect to their variance. The scales of τ are given in seconds. (From Echim et al. 2007.) (For a color version of this figure, please see the color-plate section.)

6.3.4 Solar Extreme Ultraviolet Emissions

In this subsection, ROMA is applied to an extended data set of extreme ultraviolet (EUV) images of the solar corona obtained by the SOHO spacecraft. As discussed in Section 2.4.3, this spatiotemporal data set was shown to exhibit the phenomenon of critical dynamics. Using the technique of structure function analysis, Uritsky and colleagues (2007) also demonstrated that the same data set exhibited multifractal characteristics of intermittent turbulence.

In the following text, a set of full-disk digital images of the corona taken by the extreme ultraviolet imaging telescope (EIT) on board the SOHO spacecraft in the 195Å wavelength band corresponding to the Fe XII emission at peak coronal temperatures of 1.6×10^6 °K is analyzed (Wu et al. 2011). The data included one observation period: June 29, 2001–July 28, 2001 (3,240 images at solar max with

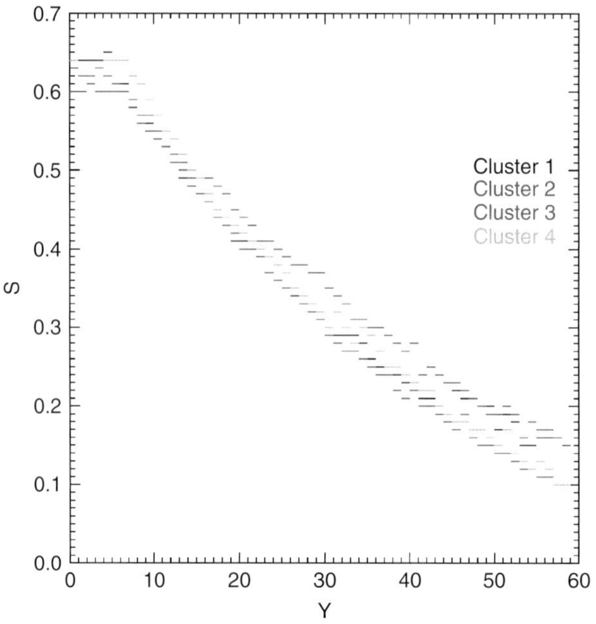

Figure 6.11. Rank-ordered multifractal spectra for the Cluster passage of Figures 6.9 and 6.10. (From Chang et al. 2010; Echim and Lamy, private communication, 2010.) (For a color version of this figure, please see the color-plate section.)

an average sunspot number of approximately 64.0) with a typical time resolution of 13.3 minutes. To reduce the optical distortions, only the central portion of the Sun disk with the linear dimensions 1040 × 1040 Mm (256 × 256 pixels with 5.6 arcsec resolution) was studied.

The dynamics of the EIT luminosity $w(t,\mathbf{r})$ captures the redistribution of the radiative flux in consecutive EIT frames due to a variety of coronal features such as loops and holes, mass ejections, and so forth. For the purpose of this study, $w(t,\mathbf{r})$ is treated as a local measure of coronal activity without any filtering of the data and thus includes various types and sizes of coronal events, Figure 6.12.

The PDFs $P(|\delta w|,\delta)$ of $\delta w(\mathbf{r})$ of one image at various scales δ are constructed, where $\delta w(\mathbf{r},\delta) = w(\mathbf{r}+\delta\mathbf{i}) - w(\mathbf{r})$ with $\mathbf{i} = \mathbf{i}_x$ or $\mathbf{i}_y$, δ is in units of 4.06 (=1040/256) Mm and $|\delta w|$ is in units of $|\delta w|_{\max}/800$, Figure 6.13. We note that the PDFs are non-Gaussian and become more and more heavy-tailed for smaller and smaller scales δ.

Figure 6.14 displays the continuum ROMA results. It is shown to include two multifractal rank-ordered regimes, one for δ from two to eight and the other for δ from twenty to eighty. In Figures 6.15 and 6.16, the functions $s(Y)$ s and $P_s(Y)$ s are shown to reproduce faithfully the observed PDFs.

This interesting result may indicate the existence of different physical mechanisms of energy dissipation in the solar corona corresponding to meso- and supergranulation scales of the underlying photospheric network. Thus this example demonstrates a double rank ordering of the multifractal nature of the turbulent intermittency. The

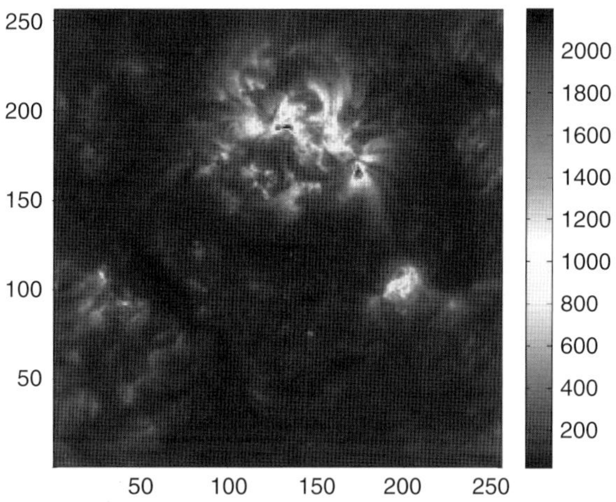

Figure 6.12. The Luminosity $w(\mathbf{r})$ in arbitrary units of a SOHO image. Only the central portion of the Sun disk with the linear dimensions 1040 × 1040 Mm (256 × 256 pixels with 5.6 arcsec resolution) was studied. (From Wu et al. 2011.) (For a color version of this figure, please see the color-plate section.)

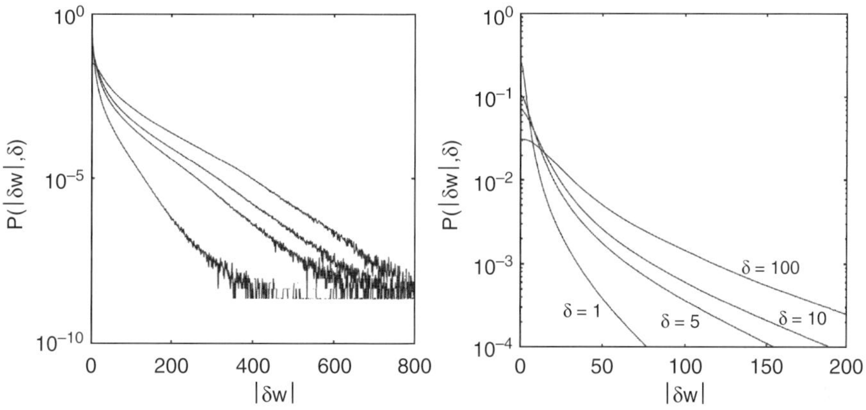

Figure 6.13. Probability distribution functions of the luminosity fluctuations $P(\delta|w(r)|,\delta)$ of a SOHO image. "δ" in units of 4.06 (=1040/256) Mm and $|\delta w|$ in units of $|\delta w|_{\max}/800$. (From Wu et al. 2011.)

first rank order differentiates the basic physical phenomena of scale regimes and the second rank order describes the local invariant property of intermittency based on the scaled sizes of each regime.

6.3.5 Intermittency of Broad Band Turbulence in the Ionosphere

Broadband extremely low-frequency (BBELF) electric field fluctuations have often been observed in the auroral zone. It has been suggested by Chang (2001a, b) that

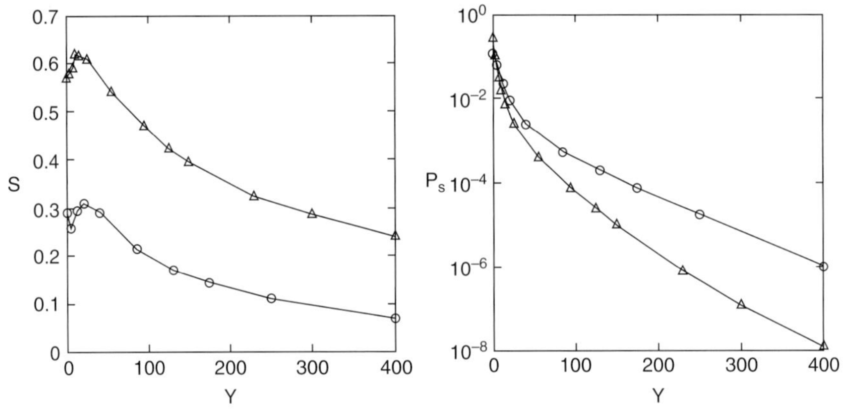

Figure 6.14. ROMA results, $s(Y)$ and $P_s(Y)$, of the PDFs of Figure 6.13. There are two multifractal rank-ordered regimes, one for δ from two to eight (triangles) and the other for δ from twenty to eighty (circles). (From Wu et al. 2011.)

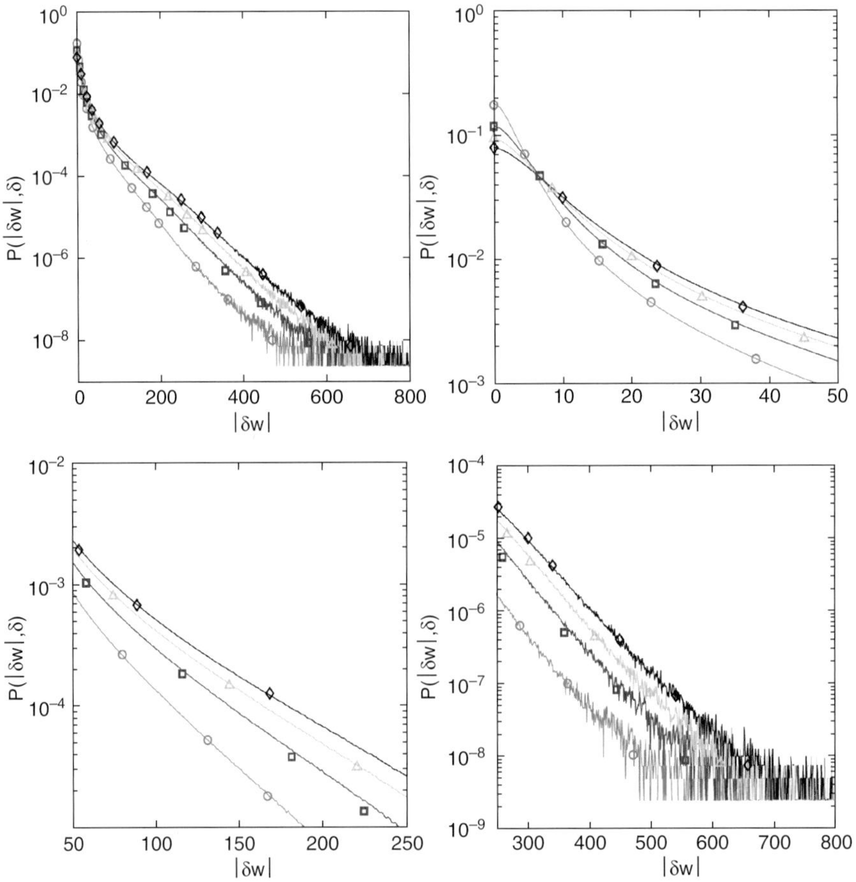

Figure 6.15. Computed PDFs $P(\delta|w(r)|,\delta)$ from ROMA relations (solid curves) and from SOHO observations (markers). δ = 2 (circles), 4 (squares), 6 (triangles), 8 (diamonds). (From Wu et al. 2011.) (For a color version of this figure, please see the color-plate section.)

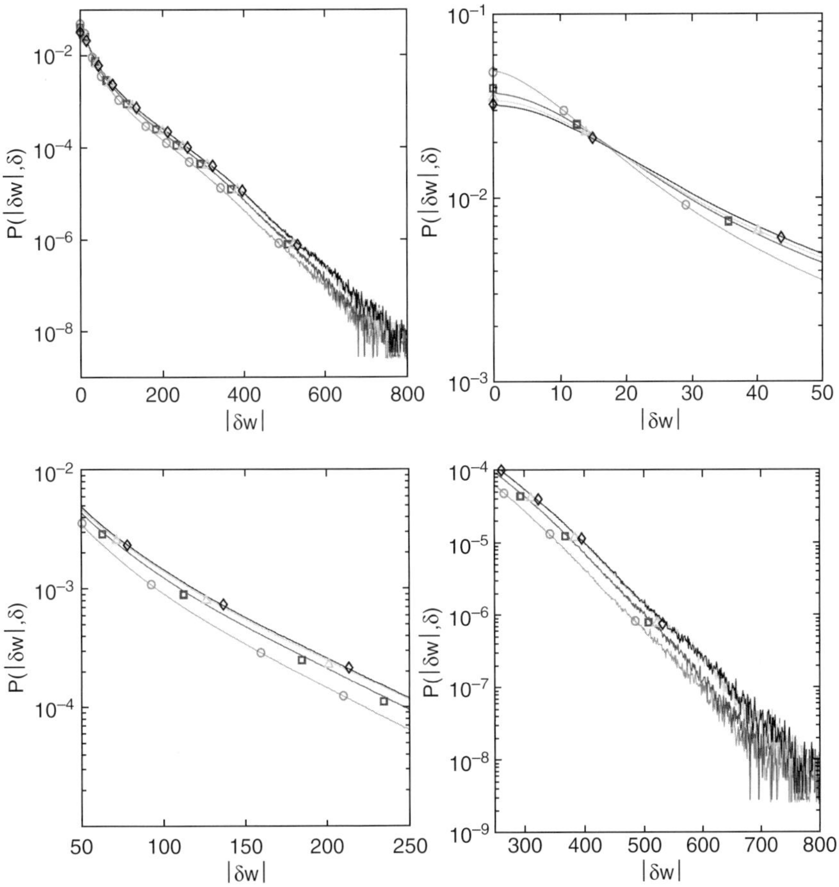

Figure 6.16. Computed PDFs $P(\delta|w(r)|,\delta)$ from ROMA relations (solid curves) and from SOHO observations (markers). δ = 20 (circles), 40 (squares), 60 (triangles), 80 (diamonds). (From Wu et al. 2011.) (For a color version of this figure, please see the color-plate section.)

such fluctuations might be the manifestation of intermittent turbulence. We shall consider in this section the multifractal characteristics of the auroral zone electric field fluctuations observed by SIERRA (Sounding of the Ion Energization Region: Resolving Ambiguities, Kintner et al. 2000). The time series of the perpendicular electric field fluctuations, as shown in Figure 6.17 (top), corresponds to where the rocket was between 550 kilometers altitude and its apogee at 755 kilometers altitude. The averaged spectral density is broadband as shown in Figure 6.17 (bottom). The PDFs of the fluctuations are non-Gaussian and strongly scale-dependent, bearing the typical intermittent characteristics, Figure 6.18. In fact, fluctuations in a subset of this time series have been shown definitively to be intermittent based on the probability distribution functions, wavelet transforms, local intermittency measures, and flatness (Tam et al. 2005).

The dominant forms of the interacting structures in the auroral zone are probably very similar to the nearly 2D oblique potential structures simulated by Seyler

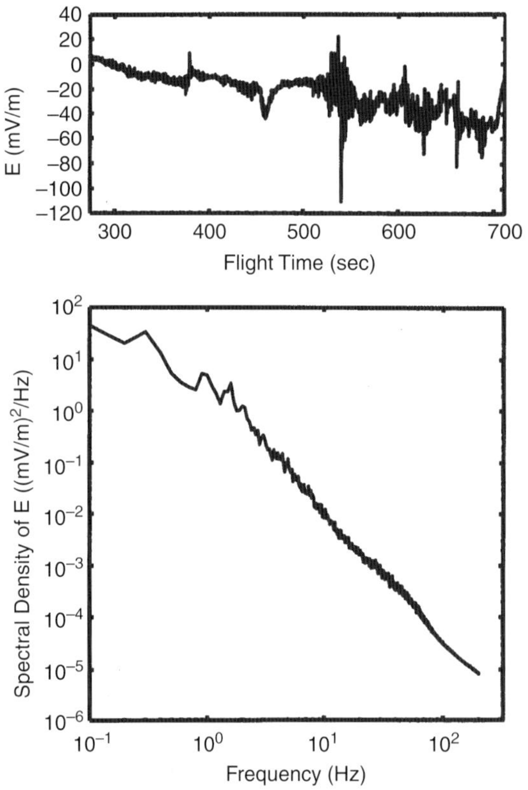

Figure 6.17. Top: Plot of the perpendicular component of the electric field E versus the flight time of SIERRA for the duration when the rocket was above 700 kilometers altitude. Bottom: Average spectral density of E over the duration, exhibiting the broadband extremely low frequency (BBELF) characteristics. (From Tam et al. 2010.)

(1990) based on the reduced MHD formulation of the inertial Alfvén fluid equations. They may be intermixed with other small-scale kinetic coherent structures (i.e., structures due to the collective effects of charged particles). The interactions of these structures are the manifestation of the localized reconnections mediated by the intermittent inertial Alfvén waves, other electromagnetic and electrostatic waves, generalized resistivity and/or coarse-grained dissipation. Thus we expect a significant fraction of the resulting fluctuations to be electrostatic, transverse, and non-resonant. And perhaps interspersed with small components of electrostatic (Bonnell et al. 1996) and/or electromagnetic waves. When detected in the spacecraft frame, the signatures of these fluctuations are Doppler shifted (Stasiewicz 2000; Temerin 1979) and have been shown to have relatively slow-moving speeds in the rest frame (Angelopoulos et al. 2001). Thus a dominant fraction of the fluctuations may be recognized as the spatial fluctuations of low frequency intermittent inertial Alfvén turbulence entrained with small-scale kinetic turbulence. And the time series and time scales may be interpreted mainly in terms of spatial scales and spatial fluctuations as described in the following text.

Assuming the horizontal speed U of the spacecraft is much larger than that of the movements of the broadband turbulent fluctuations and the geomagnetic field is essentially vertical, the time scales τ to be discussed may be viewed approximately as spatial scales $\Delta \approx U\tau$. This is (inversely) similar to the commonly invoked

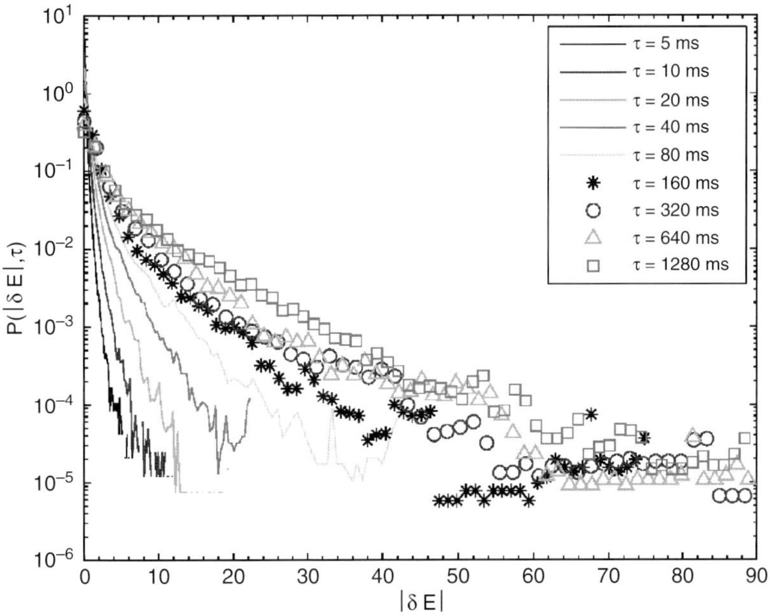

Figure 6.18. PDFs of the BBELF electric field fluctuations $|\delta E|$ in units of mV/m are non-Gaussian and strongly scale-dependent, bearing the typical intermittent characteristics. (From Tam et al. 2010.) (For a color version of this figure, please see the color-plate section.)

"Taylor hypothesis" where the speed of the solar wind is generally much faster than that of the spacecraft. Nevertheless, it must be recognized that entrained within such observed broadband turbulence there may be small fractions of electromagnetic and Doppler-shifted electrostatic propagating waves.

ROMA has been performed for these observed BBELF signals (Tam et al. 2010; Tam and Chang 2011). It may be shown as demonstrated in the following text that the observed fluctuations span across four contiguous regimes of scales with different multifractal characteristics. Figure 6.19 shows the plots of the structure function $S_q(\tau)$ versus τ at a few moment orders q for the electric field fluctuations. It is clear from the plot for $q = 1$ (top left panel) that a simple power law cannot be applied throughout the entire range of time scales. There are at least two power-law regimes with the crossover time scale appearing at $\tau \approx 80$ ms. Information based on the other panels of the figure indicates that the time scales above 80 ms should be further divided into different regimes with $\tau \approx 160$ and 320 ms being the other crossover scales. Thus, the auroral zone electric field fluctuations seem to exhibit four different regimes over the time scales up to 1,280 ms: Regime 1 from $\tau \approx 5$ to 80 ms; Regime 2 from $\tau \approx 80$ to 160 ms; Regime 3 from $\tau \approx 160$ to 320 ms; and Regime 4 for $\tau \gtrsim 320$ ms. We shall, therefore rank order the time regimes parametrically by an index $i = 1,2,3$ and 4, and study the multifractal characteristics of each regime separately.

The results of the ROMA analysis, $s_1(Y_1)$ and $P_{s1}(Y_1)$, for the time Regime $i = 1$ are given in Figures 6.20 (top left panel) and 6.21. From the spectrum, we note that

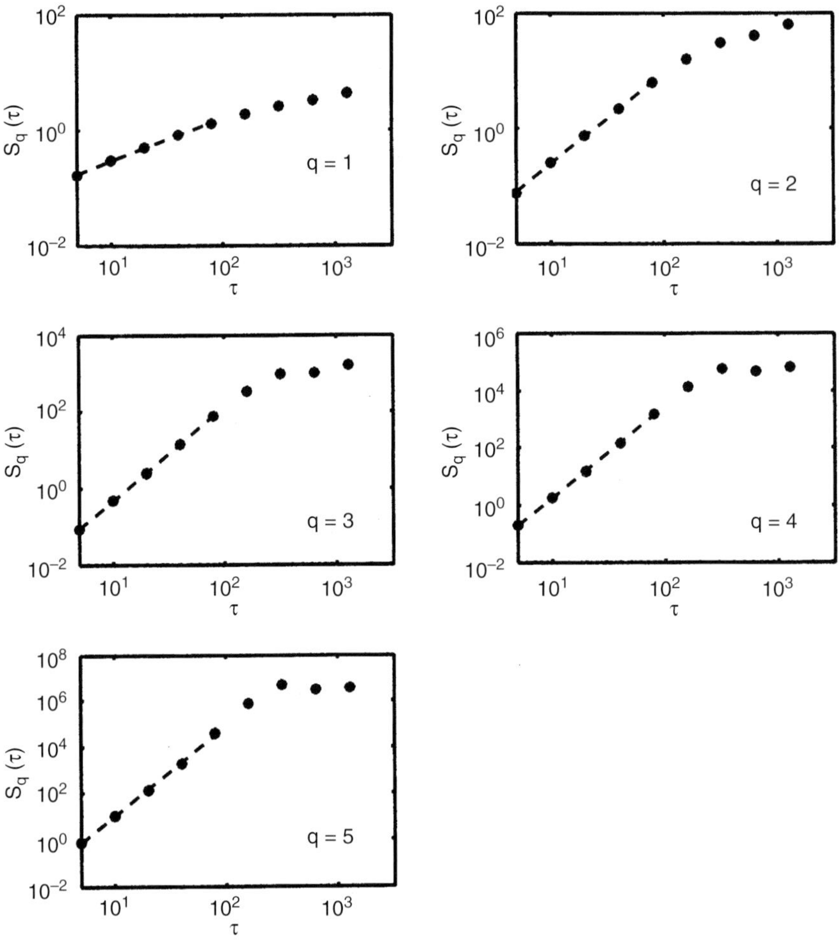

Figure 6.19. Plots of structure functions S_q of integer values ($q = 1 - 5$) versus time scale τ for the BBELF electric field fluctuations. The plots indicate the existence of four distinct regimes in time scale: 5 to 80 ms; 80 to 160 ms; 160 to 320 ms; and ≥ 320 ms. Dashed lines indicate fittings through the first time scale of regime 1. (From Tam et al. 2010.)

the fluctuations are persistent throughout the region and the persistency increases with the local scale invariant Y_1 up to a value of $s_1(Y_1)$ around 0.85 (indicating continued influence of instability in the region of smaller local scale invariants) and then settles down into a more developed turbulent state of constant persistency for larger values of the local scale invariant. And the data in this time regime collapse onto the scaled PDF $P_{s1}(Y_1)$ nicely indicating the validity of ROMA for this regime.

Similar ROMA analyses have been performed for the other time regimes, Figure 6.20. We note that the ROMA spectrum for Regime $i = 2$ has similar characteristics as that for $i = 1$, while that for Regime $i = 3$ the local scaling exponent

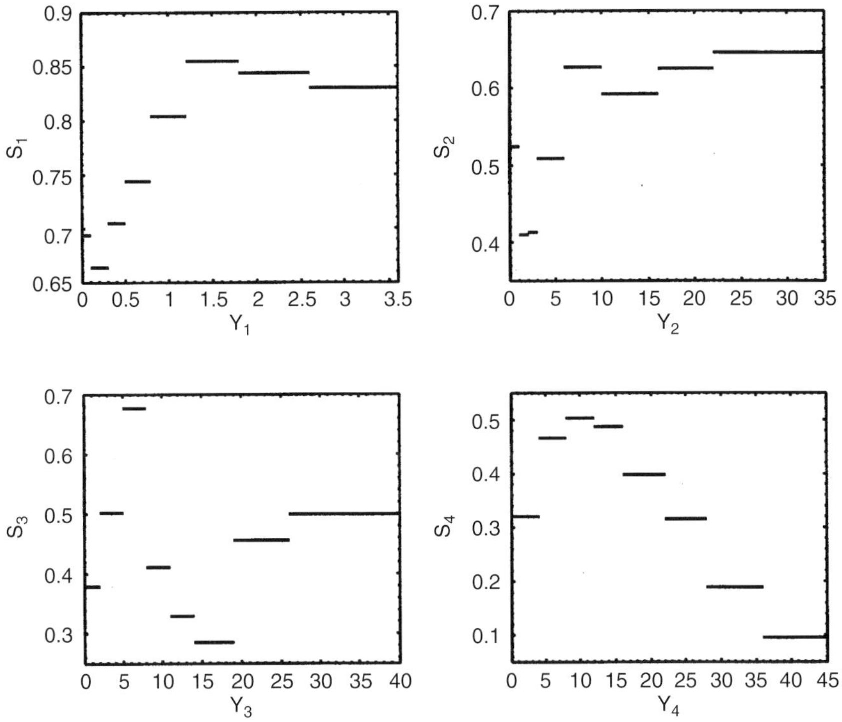

Figure 6.20. ROMA spectra $s_i(Y_i)$ for the four BBELF time regimes of the electric field fluctuations of Figure 6.18. (From Tam et al. 2010.)

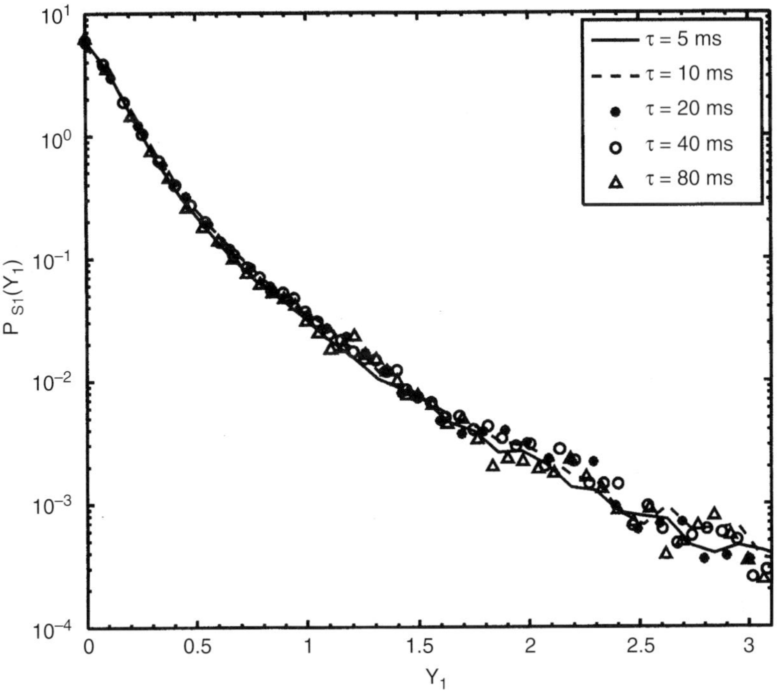

Figure 6.21. Scaled PDF $P_{s1}(Y_1)$ for the first BBELF time regime of the electric field fluctuations of Figure 6.18. (From Tam et al. 2010.)

fluctuates up and down indicating a more unstable nature in that regime. The spectrum for Regime $i = 4$ eventually settles down into fully developed turbulence for the larger local scale invariant region exhibiting general characteristics similar to that of the simulated 2D MHD turbulence discussed in the previous section. And it was found that all data collapsed nicely onto their respective ROMA-scaled PDFs as expected.

The four regimes of different scaling behavior exhibit the transition of turbulent states from the microscale, to the inertia and kinetic Alfvén scales, and eventually the MHD scales. Thus, this example represents another double rank ordering of the multifractal nature of turbulent intermittency. The first rank order differentiates the basic physical phenomena of scaling regimes and the second rank order describes the local invariant property of intermittency based on the scaled sizes of each regime.

7

Dynamic Renormalization Group and Connection between Criticality and Multifractals

7.1 Introduction

In the foregoing chapters, we provided some convincing arguments as well as numerical and observational evidences indicating that space plasma turbulence is generally in a state of dynamical complexity. To understand the phenomenon of complexity, we introduced the concepts of self-organized criticality (SOC), dynamical criticality, scaling near criticality, finite-size scaling (FSS), intermittency, multifractals, and rank ordered multifractal analysis (ROMA), leading to the basic ideas of global and local scale invariants as well as the crossover phenomena ranging over the parametric invariant regions.

It will be desirable if we may either quantitatively or qualitatively explain such invariant and crossover properties from basic physical principles. For nonlinear stochastic systems exhibiting complexity, the correlations among the fluctuations of the random dynamical fields (electric, magnetic and velocity fields, etc.) are generally very long-ranged, and there exist many correlation scales (as exhibited by the various outsized plasma coherent structures). The dynamics of such systems are notoriously difficult to handle either analytically or numerically. However, because the correlations are extremely long-ranged, it is reasonable to expect that the system will exhibit some sort of (global and/or local) invariance properties under coarse-graining scale transformations. A powerful analytical technique that exploits these invariance properties is the method of the dynamic renormalization group (DRG) which we shall briefly describe in the following text. The technique is a generalization of the static renormalization group introduced by Wilson (1971a, b) for equilibrium phase transitions (see also, Wilson and Kogut 1974). DRG was originally developed as a perturbation theory by Halperin and colleagues (1972) and first applied to hydrodynamic problems by Forster and colleagues (1977). Since then, a number of approaches toward a rational analysis of the DRG have been developed. One of such developments is the formulation of an exact closed form theory of the differential DRG by Chang and colleagues (1978). A review of the basic elements of this theory may be found in Section V of a *Physics Report* by Chang and colleagues (1992).

No matter how complex the stochastic behavior is for the turbulent system, we may conjecture that the system is somehow characterized by a set of dynamically evolving parameters $\{P_n\}$. These parameters generally are functions of the spatial and temporal variables $(\mathbf{r},t)$. We now transform the stochastic system by coarse-graining and ask how these parameters need to evolve accordingly (subject to the rules stipulated by the underlying dynamical equations; for example, those for the interacting flux tubes) such that the transformed system in a scaled sense essentially retains the characteristics of the original dynamical system. Symbolically, we may characterize this transformation of the parameters as:

$$\partial\{P_n\} / \partial\ell = R\{P_n\} \tag{7.1}$$

where R represents the coarse-graining or renormalization transformation operator and ℓ is the coarse-graining parameter. It is to be understood that the details of the transformation in general (both implicitly and explicitly) involve the spatial and temporal variables such that

$$\mathbf{r} \rightarrow \mathbf{r}' = e^{-\delta\ell}\mathbf{r} \quad \text{and} \quad t \rightarrow t' = e^{-z\delta\ell}t \tag{7.2}$$

where z is called the "dynamic exponent" and $\delta\ell$ is the incremental change of scale. The exponent "z" characterizes the asymmetric scaling nature between the spatial and temporal variables of the transformation. Such a transformation generates a group (actually a semigroup) that we shall call the DRG. The operator R is generally very complicated even for relatively simple stochastic systems. Instead of actually delving into the mathematical formalism of DRG, we shall merely explore some of the basic concepts related to such ideas, and then in the next section consider a simple example to elucidate these concepts.

Let the actual state of a dynamical system before coarse-graining be represented by $\{P_n(0)\}$. Following the prescription dictated by (7.1), the coarse-grained (i.e., renormalized) states of the system will then trace out a trajectory $\{P_n(\ell)\}$ in the phase space characterized by the parameters (see, e.g., Figure 1.3 of Chapter 1). In analogy to nonlinear dynamics, we expect that there may exist fixed points (singular points) in the phase space flow field, at which $d\{P_n(\ell)\} / d\ell = 0$ for some particular value(s) of the dynamic exponent z^*.

For most stochastic systems, fixed points are partially repulsive. And this is generally true for nonconservative systems. A partially repulsive fixed point in a two-dimensional parameter space has the characteristics of a saddle point as shown graphically in Figure 1.3. At a fixed point, the correlation length should not be changing. However, the coarse-graining transformation requires it to change with ℓ.

Therefore, to satisfy both requirements, the characteristic correlation length for the dynamic system must be either equal to infinity or zero at the fixed point. When the correlation length is infinite, the stochastic system is at criticality. Usually, a dynamical system will not be near a parametric state of a partially repulsive fixed point, for example, $\{P_n{}^*\}$ as shown in Figure 1.3. But, if the renormalization group trajectory (shown in red in Figure 1.3) leads it eventually close to it (e.g., within the green ellipse of Figure 1.3), then in coarse-grained sense the stochastic system would be close to a state of dynamic criticality. We may then approximate the coarse-graining transformation (7.1) by linearizing the operator R about this fixed point. The result is a set of coupled linear differential equations characterizing the variations of the deviations of the parameters from $\{P_n{}^*\}$.

$$d\{Q_n\} / d\ell = R_L\{Q_n\} \tag{7.3}$$

where $Q_n = P_n - P_n{}^*$ and R_L is the linearized renormalization-group operator. Let us assume that expression (7.3) may be expressed in terms of a set of eigenvalue equations:

$$d\mu_i / d\ell = \lambda_i \mu_i \tag{7.4}$$

where the eigenvectors $\mu_i{}'s$ are linear combinations of the $Q_n{}'s$ and $\lambda_i{}'s$ are the corresponding eigenvalues. (Nondiagonalizable operator R_L may render the eigenvectors incomplete and lead to logarithmic corrections to the discussion, which will not be considered here.)

The solutions of (7.4) are $\mu_i = \mu_i^0 \exp \lambda_i \ell$ where $\{\mu_i^0\}$ represents some arbitrary state that is close to $\{P_n^*\}$. From this, it can be verified immediately that

$$\mu_i / \mu_j^{\lambda_i/\lambda_j} = \text{constant} \tag{7.5}$$

that is, the $\mu_i{}'s$ form power-law invariants which are independent of the coarse-graining scale transformation ℓ.

Expressing this result in words: the mathematical consequence of the linear approximation is that, close to dynamic criticality, certain linear combinations of the perturbations of the parameters that characterize the stochastic phenomenon will correlate with each other in the form of power laws. These include the (k,ω), that is, mode number and frequency, spectra of the correlations of the various fluctuations of the dynamic field variables in space and time, and so forth. The fact that the dynamical (stochastic) system must initiate from some state so that the coarse-graining trajectory will carry it to the close proximity of a fixed point such as $\{P_n^*\}$ means that there may need to be some initial tuning of the state for the system to have such behavior. For this reason, we shall refer to such dynamical systems as systems near forced and/or self-organized criticality (FSOC) (Chang 1992), or simply dynamic criticality. The phenomenon of FSOC is spatiotemporal and the invariant scaling idea permeates throughout the time-space domain. In addition, the scaling between time and space generally is not isotropic.

An eigenvector (eigen-parameter) μ_i is "irrelevant" if the corresponding eigenvalue λ_i is less than zero because in the limit of large ℓ the value of μ_i will become negligible. The reverse is true for positive $\lambda_i's$. We shall not consider the special case "$\lambda_i = 0$" that introduces singularities of fractional powers of logarithms. If there are only a small number of relevant eigenvalues with $\lambda_i > 0$, the corresponding FSOC system has only a small number of "relevant parameters" that are needed to characterize the stochastic state, that is, the system may appear to have low dimensionality in parameter space near criticality (Chang 1992).

Sometimes, for conservative stochastic systems, the coarse-grained trajectories may evolve into the domain of an attractive fixed point (i.e., a sink singularity). Then, the renormalized stochastic systems characterized by these trajectories all approach a unique stationary state without the tuning of the parameters. And the scaling results (in particular, power-law distributions) are fully characterized by the properties of the fixed point and thus universal. Such then would be the general scenario for the phenomenon of SOC as originally proposed by Bak and colleagues (1987). The procedure to deduce the power-law distributions for attractive fixed points will necessarily be different from that for partially repulsive fixed points because there are no relevant scaling fields. We shall discuss such a procedure in Section 7.3.

As mentioned previously, apparent power-law behavior has been detected in probability distributions related to the solar flares (Einaudi and Velli 1995; Lu and Hamilton 1991; Uritsky et al. 2007), in the auroral electroject (AE) burst occurrences as a function of the AE burst strength (Consolini 1997), in the global auroral ultraviolet emissions observed by UVI (Ultraviolet Imager on the Polar spacecraft) in terms of the statistics of size and energy dissipated by the magnetospheric system (Lui et al. 2000), in the probability distributions of spatiotemporal magnetospheric disturbances as seen in the UVI images of the nighttime ionosphere (Uritsky et al. 2002a), in the probability distributions of durations of Bursty Bulk Flows (Angelopoulos et al. 1999), in the magnetic field fluctuations of the Earth's magnetotail (Hosino et al. 1994; Milovanov et al. 1996), and in luminosity fluctuations of extreme ultraviolet (EUV) emissions associated with solar flares (Uritsky et al. 2007). Similar behavior has also been obtained by numerical simulations or coupled lattice automata calculations for the distributions for solar flares (Georgoulis et al. 1998; Vlahoes et al. 1995), for dynamical models of the current sheet (Klimas et al. 2000; Takalo et al. 2001; Uritsky et al. 2005), and other applications. (This invariant behavior had been sometimes interpreted as SOC, i.e., assuming that tuning was not necessary.)

In the next section, we shall illustrate such dynamical criticality ideas based on the DRG in terms of a simple model. Then, in the final section of this chapter, we shall proceed to consider complexity phenomena beyond those of dynamic criticality and discuss how such phenomena may be understood based on generalizations beyond the discussions described in Section 7.2 in terms of local invariants, multifractals and dynamic crossovers from the point of view of ROMA and the DRG.

7.2 An Illustrative Example – One-Dimensional Modified Forest Fire Model

As explained in Section 2.4.5, The modified forest-fire model (MFFM) was suggested by Drossel and Schwabl (1992) and Mossner and colleagues (1992). It is defined on a d-dimensional lattice of linear size L. Each lattice site may be empty, or occupied by a green tree or a burning tree. After each time step, the configuration of the lattice evolves according to the following rules: (1) A green tree may grow on an empty site with probability p, and (2) a green tree becomes a burning tree if one of its neighbors is burning or (3) may catch fire spontaneously through lightning with the probability f. (4) And a site with a burning tree becomes an empty site. Periodic boundary conditions are assumed. The growth rate p is assumed to be small and the lightning rate is even smaller with $f \ll p < 1$. In addition, the growth probability is limited by the condition $p \ll T^{-1}$, where T is the time needed to burn down a large forest cluster $s_{\max}$ as to be discussed in the following text. With such assumptions, burning happens very fast such that each fire of a cluster is an isolated event, so that the only relevant parameter in the model is the ratio $\theta = f / p$, which therefore sets the scale of the average number of trees burnt after one lightning. The assumption is that the defined dynamical system becomes critical as $\theta \to 0$. This last statement has caused much controversy in recent large -scale numerical studies (particularly for $d = 2$), representing a topic of true importance in the defining concepts of complexity phenomena, and will be addressed specifically in the next section.

It was shown by Drossel and colleagues (1993) that for $d = 1$, the MFFM could be solved exactly in the asymptotic stationary state under the double limit of $f \ll p < 1$ with the classical behavior of dynamic criticality. Their results seemed to be verified by moderately large-scale ($L = 2^{20}$) simulations with $f / p = 8 / 50000$ (see also, Christensen et al. 1993). In the following we shall accept such results as claimed and follow Loreto and colleagues (1996) closely to demonstrate that the same scaling behavior for this 1D MFFM may be obtained using the so-called dynamically driven renormalization group (DDRG) that was originally developed for the study of the critical behavior of sandpiles (Pietronero et al. 1994; Vespignani et al. 1995).

We shall develop this DDRG using the simplest possible procedure by generally and successively replacing two adjacent cells at scale (k) with one coarse-grained cell at scale $(k+1)$. If at least one of the two cells at scale (k) and time step (t) is empty then we shall consider the coarse-grained cell at scale $(k+1)$ to be empty. We then ask the question how the coarse-grained cell at scale $(k+1)$ and time step $(t+1)$

can become a green tree (i.e., when both cells at step $(t+1)$ are green). The answer is that there are three ways this can happen.

(i) At time step (t) and scale (k), the left cell is green while the right cell is empty but with the probability $p^{(k)}$ of growing a green tree at time step $(t+1)$.
(ii) At time step (t) and scale (k), the right cell is green while the left cell is empty but with the probability $p^{(k)}$ of growing a green tree at time step $(t+1)$.
(iii) At time step (t) and scale (k), both left and right cells are empty but with the probability of $(p^{(k)})^2$ of both growing green trees.

Therefore, the probability for the coarse-grained empty cell at time step (t) and scale $(k+1)$ to become a renormalized green tree at time step $(t+1)$ and scale $(k+1)$ is:

$$p^{(k+1)} = W_1 p^{(k)} + W_2 p^{(k)} + W_3 (p^{(k)})^2 \tag{7.6}$$

where W_α are the relative statistical weights that we will carefully consider in the following text.

Using similar arguments, we may obtain the renormalization expression for the lightning probability by starting at scale $(k+1)$ and time step (t) with both cells being green and then assigning the renormalized cell at time step $(t+1)$ as burning if one of the two cells is burning.

$$f^{(k+1)} = (f^{(k)})^2 + 2f^{(k)}(1 - f^{(k)}) \tag{7.7}$$

We shall renormalize the dynamic system covering many coarse-graining iterations. But this will generate many additional parameters involving the various joint probabilities in the full formulation. Thus Eqs. (7.6) and (7.7) may only be considered as a truncated approximation of the full renormalization-group transformation. Within such an approximation, we shall attempt to estimate the relative statistical weights W_α from the time asymptotic stationary states of the densities ρ_i of the empty $(i = 1)$ sites, the green $(i = 2)$ trees, and the burning $(i = 3)$ trees. In moderately large-scale 1D numerical simulations, the results seemed to indicate that through the dynamic process the densities do approach some stationary asymptotic values for given values of (f, p). We shall try to estimate the asymptotic stationary values of the densities using the mean field theory. These equations were derived by Christensen and colleagues (1993) and Drossel and Schwabl (1994). The results are consistent with the discussions of the mean field theory espoused by (Vespignani and Zapperi, 1997; 1998) and explained in Section 2.3. Following Christensen and colleagues (1993), the stationary equations are:

$$\rho_1 = (1 - p)\rho_1 + \rho_3 \tag{7.8a}$$

$$\rho_2 = (1-f)(1-\rho_3)^{\bar{z}}\rho_2 + p\rho_1 \tag{7.8b}$$

$$\rho_3 = 1-\rho_1-\rho_2 \tag{7.8c}$$

where (ρ_1, ρ_2, ρ_3) are the asymptotic stationary densities of the empty sites, green trees, and burning trees. "$\bar{z}$" is the coordination number, a prescription of how fire may spread to its neighboring sites. In the present application, we follow Loreto and colleagues (1995; 1996) and assume the fire spread randomly to its neighboring sites to the extent that there are trees on these sites and $\bar{z}$ for such a random neighbor model is known to be $(2d-1)$.

We now estimate the statistical weights W_α from the probabilities of having the corresponding number of green and empty cells at scale k according to the three possibilities (i), (ii), and (iii). In terms of the asymptotic stationary mean field values of ρ_i determined by Eqs. (7.8), we have:

$$W_1 = W_2 = \rho_2/(\rho_1+2\rho_2) \quad \text{and} \quad W_3 = \rho_1/(\rho_1+2\rho_2) \tag{7.9}$$

They are assumed to be valid at any level of renormalization after the asymptotic stationary state is reached. We shall assume that at this time, the density of the burning trees is much smaller than that of the green trees (Christensen et al. 1993).

The renormalization-group equations (7.6)–(7.9) are nonlinear equations. By inspection, it is easily seen that there exists a fixed point at: $f^* = p^* = 0$, and $\rho_i{}^* = (0,1,0)$ that may exhibit all the desired scaling properties. Linearizing the equations about this fixed point and diagonalizing, in the (f, p)-subspace we find two eigenvalues. One of the eigenvalues is equal to two and therefore relevant while the other is equal to one and marginal. (We note that for the stepwise renormalization of doubling the size of the cells at each step, the dividing value for relevant and irrelevant eigenvalues is "1" as contrasted to the differential renormalization procedure where the dividing value is at "0" because the eigenvalues there were expressed as exponentials.)

We now proceed to address the physical and numerical interpretations related to the relevant eigenvalue $\lambda = 2$. We find that the fixed point is partially repulsive (in fact, a saddle point) and the relevant eigenvector points in the increasing direction of $\theta = f/p$. Thus we shall assume that within the linearization approximation of the DDRG the correlation length ξ of the green trees (which is a measure of the cluster size) varies with the relevant parameter (relevant scaling field) ϑ as a power law:

$$\xi \sim \theta^{-v} \tag{7.10}$$

where v is the correlation length exponent. For each step of renormalization, the size of ξ is reduced by half, and the value of θ is increased by λ, the relevant eigenvalue. But, from Eq. (7.10), the ratio of ξ/θ^{-v} is an invariant under renormalization, therefore, we conclude that

$$v = \log 2/\log \lambda = 1 \quad \text{for} \quad \lambda = 2 \tag{7.11}$$

We now consider the near criticality scaling expression with cutoff in analogy to those discussed in Section 2.3.2.

$$N(s) = s^{-\bar{\tau}} C_s(s / s_{\max}) \tag{7.12}$$

for the number density $N(s)$ of clusters containing s trees where $\bar{\tau}$ is the probability exponent for the cluster size, and $s_{\max} \sim \theta^{-\bar{\lambda}}$ with $\bar{\lambda}$ being the characteristic size exponent. The mean size of the cluster of trees (which is the same as the mean size of trees destroyed by a lightening stroke) is then:

$$\bar{s} = \sum_{1}^{\infty} s^2 N(s) / \sum_{'}^{\infty} sN(s) \sim s_{\max}^{3-\bar{\tau}} \tag{7.13}$$

Assuming the mean cluster radius $R(s) \sim s^{-\mu}$ with μ being its fractal dimension, we can easily deduce that $\xi \sim \vartheta^{-\bar{\lambda}/\mu}$. But, the mean size of trees destroyed by a lightning stroke may also be estimated as $\bar{s} = p\rho_1 / f\rho_2 \sim \theta^{-1}$ assuming that the mean densities approach constant values near criticality. Therefore,

$$\bar{\lambda} = \mu v = 1/(3-\bar{\tau}) \tag{7.14}$$

For $d = 1$, the dynamic clustering of green trees is compact. Thus, $\mu = 1$ and $\bar{\lambda} = v$ or $\bar{\tau} = 2$. As mentioned by Drossel and Schwabl (1992) and Clar and colleagues (1994), for $\bar{\tau} = 2$ the cutoff function needs to be multiplied by a logarithmic term $\ln^{-1}(s_{\max})$ to insure a finite mean forest density in the limit of $\vartheta \to 0$.

The dynamic exponent z may trivially be surmised by considering the dynamics of fire spreading. It is obvious that in the asymptotic states of $k \to \infty$ it takes two burning processes at length scale k to have a fire at length scale $k+1$. Because the spatial scale is doubled from k to $k+1$ in this time scale, we have immediately $z = 1$.

The correlation length exponent, the dynamic exponent, and the cluster probability exponent based on the preceding DDRG calculations are the same as those given by the exact theory of Drossel and colleagues (1993).

In performing the sample DDRG calculations for the 1D MFFM, we have assumed that there is an asymptotic stationary state and this seemed to be verified by moderately large-scale simulations. For $d = 2$, however, Grassberger (2003) and Pruessner and Jensen (2003; 2004) have indicated that for very large-scale numerical simulations the MFFM may not be critical and in fact that the apparent asymptotic stationary state obtained by moderately sized numerical simulations may actually be a transient state. We shall consider such discussions and other complications leading to the general complexity behavior in the next section from the point of view of the DRG.

7.3 Beyond Criticality

7.3.1 Introduction

As indicated in the previous chapters, dynamic complexity is not just a phenomenon of criticality. It is therefore of interest to inquire if the DRG may be employed to study complexity phenomena that are distinctly more complicated than simple criticality. Power-law scaling generally comes about when the DRG trajectory is attracted to the vicinity of a fixed point. What happens if the trajectory is within a domain influenced by more than one single fixed point or more generally if the transformation underlying the DRG is fully nonlinear. The global invariants of the group under such situations (if they exist) are generally not power laws. Nevertheless, as we shall argue, it may still be possible to talk about local invariants that are power laws with the nonlinearity of transformations prescribing the specific phenomena as crossovers. It is with such a concept in mind that we may provide a connection between the properties of dynamic criticality and general complexity.

7.3.2 Generalized Near Criticality and Finite-Size Scaling Laws, Complexity, and Multifractals

Let us consider first the concept of near criticality scaling laws in dynamic phenomena. As described several times in the previous chapters, such scaling is the general assumed property of self-similarity for systems obeying dynamic criticality. It may be understood from the point of view of generalized homogeneous functions (GHFs) or the linearized DRG near fixed points.

In critical dynamics, the probability distribution function (PDF) is generally assumed to be independent of the threshold cutoff and for a finite system it is also characterized by a cutoff scaling function. For example, for the sandpile model, we defined in Chapter 2 the probability distribution function $P_s(s,\varepsilon)$ of active sites near criticality and postulated a scaling relation reproduced as follows:

$$P_s(s,\varepsilon) = s^{-\tau_s} C_s(s / s_c(\varepsilon)) \tag{2.11a}$$

where C_s was the cutoff or rescaling function, $s_c \sim \varepsilon^{-\nu_s}$ was the characteristic avalanche size, τ_s was the site probability exponent, ν_s was the corresponding characteristic size exponent, and ε was the dissipation parameter.

Generalizing, we may replace in (2.11) the dissipative parameter ε by a control parameter (or relevant scaling field, θ). In the expression, $I = s / s_c$ is a global invariant of the critical phenomenon under consideration. Because s_c is related to the relevant scaling field (and therefore the correlation length for critical systems describable by a partially repulsive fixed point in terms of the renormalization-group language), it may be shown that $s_c \sim \xi^{\nu_s/\nu}$. Such a near criticality scaling relation may be applied to a system of finite size. Because the correlation length approaches the linear finite scale (i.e., $\xi \sim L$) near criticality with $\xi \sim \theta^{-v}$, we obtain the FSS law

$$\bar{P}_s(s,L) = L^{-v_s \tau_s / v} \bar{C}_s(s / L^{v_s / v}) \tag{7.15}$$

This FSS expression (7.15) contains two independent scaling exponents: $\tilde{\tau} \equiv \tau_s \tilde{v}$ and $\tilde{v} \equiv v_s / v$. One may attempt to map the probabilities for $s < \xi$ onto one power law with the exponent $\tilde{\tau}$ and leave the other exponent $\tilde{v}$ as the FSS exponent. If FSS is satisfied, $\tilde{v}$ is a constant. Otherwise, $\tilde{v}$ varies with the local invariant $Y = s / L^{\tilde{v}}$ leading to an implicit ROMA spectrum $\tilde{v}(Y)$ for the generalized FSS expression of the scaling function $\bar{C}_s(Y)$, which may have an arbitrary shape or the usually assumed smooth cutoff shape.

Alternatively, if we remove the threshold requirement for the probabilities, we may view the expression (7.15) as the monofractal scaling form of the PDFs $\bar{P}(s,L)$ for different L s. In that case, the exponent τ_s with the requirement that the PDFs are normalized must be equal to one.

Expression (7.15) with $\tau_s = 1$ is analogous to the scale-dependent probability distribution function PDF $P(X,\delta)$ of incremental changes X that exhibited self-similar behavior and satisfied the one-parameter scaling form as discussed in Chapters 4 and 6:

$$P(X,\delta)\delta^s = P_s(X / \delta^s) \tag{4.14}$$

where s is the Hurst (or monofractal scaling exponent) and δ is the incremental scale. Thus all PDFs essentially collapse onto one master scaling function P_s.

When $P(X,\delta)$ did not scale according to (4.14), we developed the ROMA formalism where we considered the invariant X / δ^s as a local invariant that changed as incremental scale size exponent s (now a local Hurst exponent) changed. Thus, in this generalized scenario of scaling, the global invariant was no longer a power law.

We may now extend the FSS expression (7.15) with $\tilde{\tau} = 1$ the same way. When the dynamic behavior does not exactly fit the classical critical scaling idea, FSS may be generalized using the ROMA formalism to describe the data with L taking on the role of δ. In fact, ROMA is now applied to the full range of the PDFs without incorporating the threshold cutoffs.

In terms of DRG, the scaling exponents of the GHFs are no longer constants. They vary as the renormalization group (RG) scaling parameter ℓ varies. In other words, the RG transformations are no longer linear and the renormalized dynamic system is no longer close to a particular fixed point. We shall now demonstrate this through a simple toy model example in the next subsection and relate it to the implicit ROMA spectrum $\tilde{v}(Y)$ as described. (Examples of such nonlinear crossover effects have been worked out in detail for several interesting exact and approximate examples of static and dynamic renormalized systems and thoroughly reviewed in a comprehensive monograph by Chang et al. [1992]. These discussions, the associated mathematics, and related formalisms are, however, beyond the scope of this introductory treatise on complexity.)

7.3.3 A Toy Model of Global Crossover among Multiple Fixed Points

In Sections 2.4.5 and 7.2, we briefly considered the revised forest-fire model (RFFM) of Consolini and De Michelis (2001) and the DDRG analysis (Loreto et al. 1996) of the 1D MFFM of Drossel and Schwabl (1992). It was pointed out within these discussions that very large-scale 2D numerical simulations (Grassberger 2002; Prussner and Jensen 2002; 2004) for the MFFM seemed to cast doubt on the earlier simulation results for moderate-sized lattices. These calculations indicated that the model was not critical for very small values of the relevant parameter and that the density of trees for large CPU times seemed to be much smaller than those given by previous estimates. In fact, the long time behavior did not appear to indicate the existence of an asymptotic stationary state.

Yet, by assuming that the MFFM did have an asymptotic stationary state, Loreto and colleagues (1995, 1996) seemed to have obtained the 2D DDRG results that were comparable to those obtained with moderately sized numerical simulations.

We shall demonstrate in this subsection (Chang et al. 2003; Tam et al. 2000) that a 2D toy model based on a variation of the Loreto and colleagues model and with an assumed asymptotic stationary state will be able to provide some interesting general comments as well as useful insights related to the concepts of fixed points of various characteristics and nonlinear crossover behavior among such fixed points based on the nonlinear DRG formalism.

The considered toy 2D DDRG MFFM model is constructed very similar to the 1D DDRG of Section 7.2 and based on the coarse graining of doubling the spatial scale in both directions. The coarse-graining spanning rules are depicted graphically in Figure 7.1. Taking into account of the relative statistical weights based on the asymptotic stationary density states, the recursion relations for (f, p) with such a spanning and iterative scheme are given below:

$$p' = \frac{\rho_1^2(4p^3 - 3p^4) + 4\rho_1\rho_2(1-f)(3p^2 - 2p^3) + 6\rho_2^2(1-f)^2(2p - p^2)}{\rho_1^2 + 4\rho_1\rho_2 + 6\rho_2^2} \tag{7.16}$$

$$f' = \frac{4\rho_1[1-(1-f)^3] + \rho_2[1-(1-f)^4]}{4\rho_1 + \rho_2} \tag{7.17}$$

where the prime's over (f', p') denote the values after the renormalized step and the ρ_i's are the stationary mean values of the densities satisfying Eqs. (7.8) as explained in Section 7.2 and displayed here graphically in Figure 7.2. Here we choose the coordination number "$z = 4$," the total number of available neighboring sites on a square lattice.

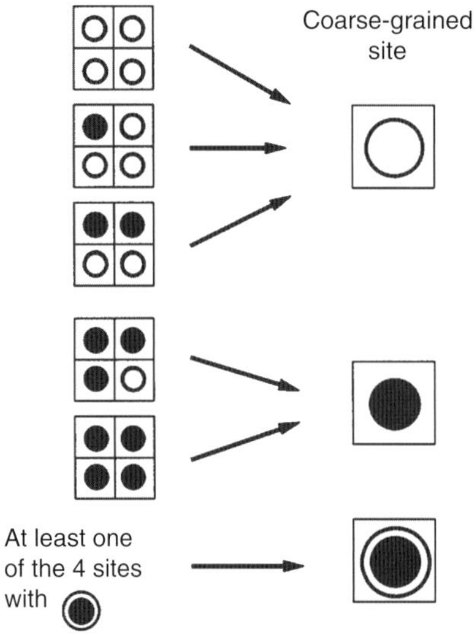

Figure 7.1. Coarse-graining spanning rules for the toy 2D DDRG MFFM. Circle (empty site), filled circle (green tree), and partially filled circle (burning tree).

These DDRG equations are slightly different from those of Loreto and colleagues (1995; 1996). The difference is not expected to affect the ensuing discussions in this subsection, which should be interpreted with the general nonlinear crossover phenomenon in mind and not constrained by the limitations of MFFM.

There are four fixed points for this toy model within the ranges of f and p between 0 and 1. They are at (f^*, p^*) =$(0,0), (0, \sim 0.282), (1,0), (0,1)$. The first one is a saddle point with $\rho^* = (3/4, 1/4, 0)$ and the second is a physically realizable but unstable source singularity with $\rho^* \approx (0.535, 0.314, 0.151)$. The last two fixed points with $\rho^* = (1,0,0)$ and $\rho^* \approx (0.302, 0.396, 0.302)$, respectively, are not physically realizable from the point of view of MFFM. Nevertheless, they exert important influence on the behavior of the RG trajectories in the (f, p) – plane as can be seen in Figure 7.3.

We note from Figure 7.3 that there are three different types of RG trajectories to the lower left of the separatrix connecting the fixed points at (0,1) and (1,0). A trajectory of the first type initiates from some locality away from and to the left of the source fixed point $(0, \sim 0.282)$. It traverses toward the partially repulsive fixed point (0,0) before turning quite abruptly toward the nonphysical sink fixed point (1,0). Such a trajectory characterizes the critical scaling behavior of the MFFM of the numerical results from the early 1990s. When it traverses close to the vicinity of the partially repulsive fixed point the correlation length exponent ν is dictated by the relevant eigenvalue λ evaluated at (0,0). The calculated values for the RG are $\lambda \approx 3.077$ and $\nu = \log 2 / \log \lambda \approx 0.617$, and the relevant scaling field is $\theta = f / p$.

Assuming the cluster of green trees is compact, which is consistent with the choice of the coordination number of $\overline{z} = 4$, we have the fractal dimension $\mu = 2$. Thus, from (Eq. 7.14 of Section 7.2), the scaling relation near criticality, we have $2\nu = 1/(3 - \overline{\tau})$, which yields the cluster-size probability exponent of $\overline{\tau} \approx 2.189$.

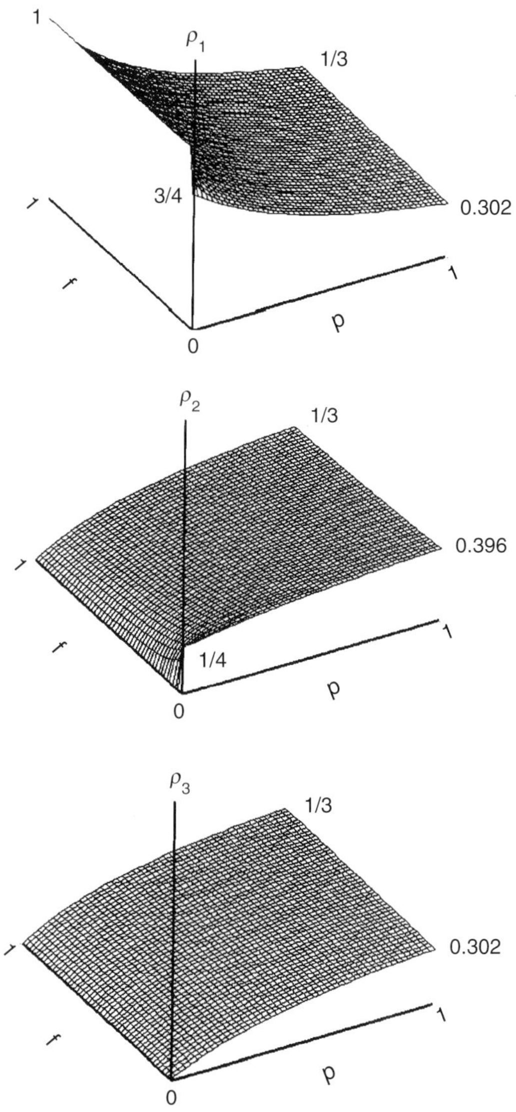

Figure 7.2. Stationary mean field density profiles for the toy 2D DDRG MFFM.

Both the preceding values of the correlation length exponent and the size probability exponent are consistent with the results obtained based on the early moderate-sized simulation results of MFFM.

As such a trajectory traverses away from the fixed point (0,0) toward the nonphysical fixed point (1,0), one expects the correlation length exponent to first decrease toward the mean field value of 0.5 and then continue to decrease as it approaches (1,0) because the nonphysical fixed point is a sink singularity with irrelevant eigenvalues.

However, as we recall from our previous discussions of MFFM, the value of f must be much smaller than the value of p. Thus, as such a trajectory skirts around the partially repulsive fixed point and begins to traverse toward the nonphysical

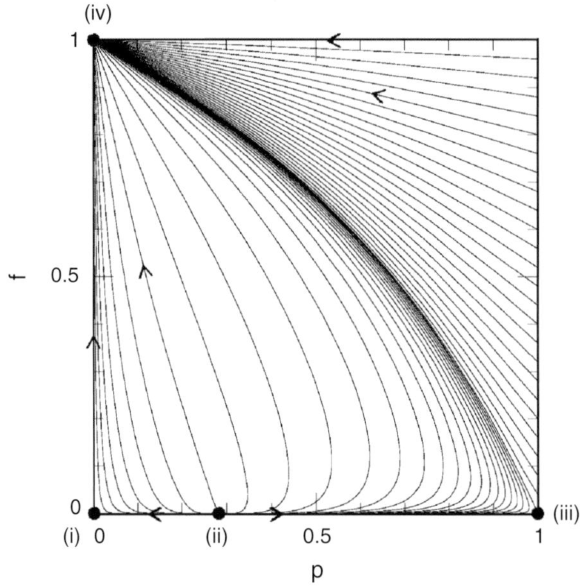

Figure 7.3. Renormalization group trajectories for the toy 2D DDRG MFFM.

fixed point (1,0), the preceding restriction will soon be violated. This is perhaps the reason why very large-scale simulations at very small values of $\vartheta = f / p$ produced inconclusive and apparently noncritical results, leading to the indication that the assumed asymptotic stationary state of the dynamic system might only actually be a transient behavior.

A trajectory of the second type initiates from the source fixed point (0, ~0.282) or its vicinity and migrates directly toward the nonphysical fixed point (1,0). Because of the nature of the source singularity, there are two relevant eigenvalues at and near this fixed point. The correlation length exponent will be given by the larger λ_1 of the relevant eigenvalues. At this source fixed point, $\lambda_1 \approx 3.128$, yielding a correlation length exponent of $\nu \approx 0.608$. (And, assuming compactness as well as the scaling relation with cutoff near criticality, we can calculate the corresponding size probability exponent $\bar{\tau} \approx 2.20$.) We note that the relevant scaling field θ is now expressed in terms of the ratio of the values of f and $p - p^*$, where $p^* \sim 0.282$.

A local crossover exponent may be defined as the ratio of the two eigenvalues, λ_1 and λ_2. It characterizes the competition between the scaling behavior along the strong RG path direction and that of the weak non-RG path directions (Chang 1992; Chang et al. 1973).

The difference between the nature of the evolution of the eigenvalues and exponents such as λ and $\bar{\tau}$ for trajectories of the second type from that of the first type is that it is not influenced by the saddle fixed point at (0,0). Both eigenvalues along such a trajectory will become irrelevant as the nonphysical fixed point with the sink singularity is approached. At that point, the assumption of $f / p \ll 1$ will have long been violated. Dynamic criticality generally can only occur quite transiently for states along a trajectory of the second type because the fixed point is a source singularity and therefore unstable. Only dynamical states that originate close to such a fixed point may have a transient critical behavior and the coarse-grained states soon crossover to other local invariant states as to be explained.

Finally, trajectories of the third type are those initiating from the right of the source fixed point (0, ~0.282) and affected by the partially repulsive fixed point (0,1) before being attracted by the nonphysical fixed point (1,0). These trajectories will have properties similar in parts to those discussed for the first two types. However, its physical validity will be limited by the value of p as it traverses toward the fixed point (0,1), which must be smaller than 1 according to the MFFM concept.

We may estimate the variation of the cluster size probability distribution exponent $\bar{\tau}$ along a RG trajectory from the local mean field stationary density of green trees, ρ_2, approximately (Loreto et al. 1995; 1996). With the assumption of compactness, we estimate the cluster size $s \sim r^2$ where r is the average radius of the cluster. Then, the probability K_k that an active process of catching fire is limited between the renormalization steps of block size $b(k)$ and size $b(k+1) = 2b(k)$ is

$$K_k = \int_{b(k)}^{b(k+1)} P(r)dr \Big/ \int_{b(k)}^{\infty} P(r)dr \tag{7.18}$$

From Eq. (7.12), $P(s) = sN(s) \sim s^{1-\bar{\tau}} C_s(s / s_{\max})$ where C_s is the cutoff function. Thus, $P(r) \sim 2rP(s) \sim r^{3-2\bar{\tau}}$ and from (7.18) we obtain

$$K_k \approx 1 - 2^{2(2-\bar{\tau}_k)} \tag{7.19}$$

And, therefore,

$$\tau_k \cong 2 - \log(1 - K_k)/2\log 2 \tag{7.20}$$

As pointed out by Drossel and Schwabl (1992) for a trajectory of type 1, “2” is the lower bound for $\bar{\tau}$, which corresponds to $K_k = 0$. And if $\bar{\tau} = 2$, then $N(s)$ must contain a logarithmic correction $\ln^{-1}(s_{\max})$ to guarantee a finite mean forest density in the limit of vanishing scaling variable(s).

The value of K_k may be estimated as $K_k \approx (1 - \rho_2)^4$ where ρ_2 is the mean field stationary density of green trees at the local parametric values of (f, p) at step k along the RG path, that is, it is the probability that at the local renormalization step with all the nearest neighbors of a burning tree are empty (again, assuming compactness and the existence of an asymptotic stationary state).

We note that the preceding method of determining the power-law distribution of avalanche size does not depend on the concept of deviations of the relevant scaling field from criticality as characterized by partially repulsive points. Thus it is equally useful for the determination of exponents of power-law distributions for attractive fixed points that characterize SOC states of Bak and colleagues (1987).

The value of $\bar{\tau}_k$ increases continuously for the first type of trajectories as the RG step k increases. The reason is that along such a trajectory the value of p decreases and that of f increases. Therefore, accordingly the mean stationary density of green trees ρ_2 decreases continuously, leading to the increase of the values of K_k and $\bar{\tau}_k$ with k. We may evaluate the local values of $\bar{\tau}_k$ for trajectories of type 2 in a similar procedure.

We point out here that the exponents and their variations actually will be different for each separate trajectory of any of the three types. In fact, it is the local values of the parameters (f, p) that determine the values of the local exponents.

Near either the partially repulsive fixed point (0,0) or the source fixed point (0,~0.282), there will be a domain of influence within which the scaling relation with cutoff

$$P(s) = sN(s) \sim s^{1-\bar{\tau}} C_s(s / s_{\max}), \quad (s_{\max} \sim \theta^{-\bar{\lambda}}) \tag{7.21}$$

is approximately satisfied in the global sense and, thus, the concept of FSOC or dynamic criticality may be applied although we must note that the source fixed point is unstable, implying that the critical behavior is quite transient there. (We also note here that the definition of θ varies among the different types of trajectories.)

Along each trajectory of RG beyond such domain, the scaling exponents can change (sometimes quite drastically) and the definition(s) of scaling directions vary among the different types of trajectories, but nevertheless the scaling relation with cutoff (7.21) is satisfied locally in terms of the local exponents. The concept of ROMA must then be considered as the local scaling exponents crossover from one renormalization step to the next. (The procedure is similar to that used to describe the complexity data with the finite length L or its equivalent that took on the role of the incremental scale size δ for intermittent fluctuations as explained in detail in Section 7.3.2.)

In fact, there generally may exist a vast range of scaling variables (f and p for this toy example) where the characteristics of the power-law behavior, the fractal dimension, the scaling direction(s), and the size distribution exponent can change significantly from one locality to another, leading to the necessity of understanding the stochastic behavior in terms of multiple parameter ROMA scaling ideas such as those discussed in Chapter 6.

(We emphasize that general stochastic processes exhibiting complexity beyond the limitations of this toy model can encompass any domain within the phase space of their scaling parameters and therefore must be understood in terms of the crossover of local invariants of any realizable RG steps within the phase space as long as it is physically meaningful. Because these are local invariance concepts, we may talk about rigorous local power-law scaling instead of merely power-law scaling near singularities.)

By choice, we considered a simple toy model that required the minimum of mathematics and physical concepts, which provided a sufficiently general qualitative picture of how multifractal behavior might be understood in a nonlinear DRG dialogue. The example delineated clearly the difference and interrelationship between global FSS in critical dynamics and ROMA and its generalizations associated with intermittent, multiparameter, multifractal processes. As emphasized before, for readers who are motivated by this introductory text and interested in studying explicit analytical examples of such nonlinear crossover behavior and processes, they are encouraged to peruse the contents of the ideas espoused in the monograph on the subject by Chang and colleagues (1992).

Finale

This completes our abbreviated tour of criticality, SOC, FSOC, scaling near criticality, FSS, intermittency, multifractals, ROMA, and the dynamic renormalization group, with applications to selected examples related to the physics of space plasmas. It is my sincere hope that the readers will be motivated to apply the concepts and methodologies espoused in this introductory monograph to their own research activities. There is much to be explored and studied. Have fun and enjoy your visit to this wonderland of complexity!

References

Alexandrova, O., Solar wind vs. magnetosheath turbulence and Alfvén vortices, *Nonlinear Processes in Geophysics* 15, p. 95, 2008.

Alexandrova, O., and J. Saur, Alfvén vortices in Saturn's magnetosheath: Cassini observations, *Geophys. Res. Lett.* 35, doi:10/1029/2008/GL034411, 2008.

Alexandrova, O., A. Mangeney, M. Maksimovic, N. Cornilleau-Wehrlin, J. M. Bosqued, and M. André, Alfvén vortex filaments observed in magnetosheath downstream of a quasi-perpendicular bow shock, *J. Geophys. Res.* 111, A12208, doi:10.1029/2006JA011934, 2006.

Angelopoulos, V., T. Mukai, and S. Kokubun, Evidence for intermittency in Earth's
plasma sheet and implications for self-organized criticality, *Phys. Plasmas* 6, p. 4161, 1999.

Angelopoulos, V., F. S. Mozer, J. Bonnell, M. Temerin, M. Somoza, W. K. Peterson, H. L. Collin, and B. Giles, Wave power studies of cusp crossings with the Polar satellite, *J. Geophys. Res.* 106, p. 5987, 2001.

Arimitsu, T., and N. Arimitsu, Analysis of fully developed turbulence by a generalized statistics, *Prog. Theor. Phys.* 105, p. 355, 2001.

Arimitsu, T., and N. Arimitsu, Analysis of fully developed turbulence in terms of Tsallis statistics, *Phys. Rev. E* 61, p. 3237, 2000a

Arimitsu, T., and N. Arimitsu, Tsallis statistics and fully developed turbulence, *J. Phys. A* (Math. Gen) 33, L235, 2000b.

Aschwanden, M. J., ed., *Self-organized Criticality Systems*, Open Academic Press, Berlin and Warsaw, 2013.

Bacry, E., J. F. Muzy, and A. Arnéodo, Singularity spectrum of fractal signals from wavelet analysis: exact results, *J. Stat. Phys.* 70, p. 635, 1993.

Bak, P., K. Chen, and C. Tang, A forest-fire model and some thoughts on turbulence, *Phys. Lett. A* 147, p.297, 1990.

Bak, P., C. Tang, and K. Wiesenfeld, Self-organized criticality: an explanation of 1/f noise. *Phys. Rev. Lett.* 59, p. 381, 1987.

Bak, P., C. Tang, and K. Wiesenfeld, Self-organized criticality, *Phys. Rev. A* 38, p. 364, 1988.

Baker, D. N., A. J. Klimas, R. L. McPherson, and J. Büchner, The evolution from weak to strong geomagnetic activity: An interpretation in terms of deterministic chaos. *Geophys. Res. Lett.* 17, p. 41, 1990.

Beck, C., Application of generalized thermostatistics to fully developed turbulence, *Physica A* 277, p. 115, 2000.

Beck, C., and F. Schlögl, *Thermodynamics of Chaotic Systems: An Introduction*, Cambridge University Press, Cambridge, 1993.

Becker, T., H. Devries, and B. Eckhardt, Dynamics of a stochastically driven running sandpile, *J. Nonlinear Sci.* 5, p. 167, 1995.

Benzi, R., S. Ciliberto, R. Tripiccione, C. Baudet, F. Massaioli, and S. Succi, Extended self-similarity in turbulent flows, *Phys. Rev. E* 48, p. R29, 1993.

Biskamp, D., *Nonlinear Magnetohydrodynamics*, Cambridge University Press, Cambridge, 1993.

Biskamp, D., *Magnetohydrodynamic Turbulence*, Cambridge University Press, Cambridge, 2003.

Bonnell, J., P. Kintner, J.-E. Wahlund, K. Lynch, and R. Arnoldy, Interferometric determination of broadband ELF wave phase velocity within a region of transverse auroral ion acceleration, *Geophys. Res. Lett.* 23, p. 3297, 1996.

Bruno, R., B. Bavassano, E. Pietropaolo, V. Carbone, and P. Veltri P, Effects of intermittency on interplanetary velocity and magnetic field fluctuations anisotropy, *Geophys. Res. Lett.* 26, p. 3185, 1999.

Bruno, R., V. Carbone, P. Veltri, E. Pietropaolo, and B. Bavassano, Identifying intermittency events in the solar wind, *Planetary Space Sci.* 49, p. 1201, 2001.

Bruno, R., V. Carbone, S. Chapman, B. Hnat, A. Noullez, and L. Sorriso-Valvo, Intermittent character of interplanetary magnetic field fluctuations, *Phys. Plasmas* 14, 032901, 2007.

Burlaga, L. F., Intermittent turbulence in the solar wind, *J. Geophys. Res.* 96, p. 5847, 1991a.

Burlaga, L. F., Multifractal structure of the interplanetary magnetic field: Voyager 2 observations near 25 AU. 1987–1988, *Geophys. Res. Lett.* 18, p. 69, 1991b.

Burlaga, L. F., Multifractal structure of speed fluctuations in recurrent streams at 1AU and near 6 AU, *Geophys. Res. Lett.* 18, p. 1651, 1991c.

Burlaga, L. F., Intermittent turbulence in large-scale velocity fluctuations at 1 AU near solar maximum, *J. Geophys. R.* 98, p. 17467, 1993.

Carbone, V., Cascade model for intermittency in fully developed magnetohydrodynamic turbulence, *Phys. Rev. Lett.* 71, p. 1546, 1993.

Carbone, V., Scaling exponents of the velocity structure functions in the interplanetary medium. *Ann. Geophys.* 12, p. 585, 1994.

Carbone V., and R. Bruno, Cancellation exponents and multifractal scaling laws in the solar wind magnetohydrodynamic turbulence, *Ann. Geophys.* 14, p. 777, 1996.

Carbone, V., R. Bruno, and P. Veltri, Evidence for extended self-similarity in hydromagnetic turbulence. *Geophys. Res. Lett.* 23, p. 121, 1996.

Carbone, V., P. Veltri, and R. Bruno, Experimental evidence for differences in the extended self-similarity scaling laws between fluid and magnetohydrodynamic turbulent flows, *Phys. Rev. Lett.* 75, p. 3110, 1995.

Castaing, B., Y. Gagne, and E. J. Hopfinger, Velocity probability density functions of high Reynolds number turbulence, *Physica D* 46, p. 177, 1990.

Chang, T., Low-dimensional behavior and symmetry breaking of stochastic systems near criticality – Can these effects be observed in space and in the laboratory? *IEEE Trans. Plasma Science* 20, p. 691, 1992.

Chang, T., Sporadic localized reconnections and multiscale intermittent turbulence in the magnetotail. In *Geospace Mass and Energy Flow*, edited by J. L. Horwitz, D. I. Gallagher, and W. K. Peterson, Geophysical Monograph 104, AGU, Washington, DC, p. 193, 1998.

Chang, T., Self-organized criticality, multifractal spectra, sporadic localized reconnections and intermittent turbulence in the magnetotail, *Physics of Plasmas* 6, p. 4137, 1999.

Chang, T., An example of resonances, coherent structures and topological phase transitions – the origin of the low frequency broadband spectrum in the auroral zone, *Nonlinear Processes in Geophysics* 8, p. 175, 2001a.

Chang, T., Colloid-like behavior and topological phase transitions in space plasmas: intermittent low frequency turbulence in the auroral zone, *Physica Scripta* T89, p. 80, 2001b.

Chang, T., Dynamical complexity in space plasmas, in *Encyclopedia of Complexity and Systems Science*, Editor in Chief, R. A. Meyers, doi: 10.1007/978-0-387-30440-3_510, Springer, New York, 2009.

Chang, T., and C. C. Wu, Dynamical complexity, intermittent turbulence, coarse-grained dissipation, criticality and multifractal processes, in *Turbulence and Nonlinear Processes in Astrophysical Plasmas*, edited by D. Shaikh and G. Zank, AIP Proc, American Institute of Physics, Melville, NY, 932, p. 161, 2007.

Chang, T., and C. C. Wu, Rank-ordered multifractal spectrum for intermittent fluctuations, *Phys. Rev. E* 77, 045401(R), doi:10.1103/PhysRevE.77.045401, 2008.

Chang, T., A. Hankey, and H. E. Stanley, Generalized scaling hypothesis in multicomponent systems. 1. Classification of critical points by order and scaling at tricriical points, *Phys. Rev. B* 8, p. 346, 1973.

Chang, T., J. F. Nicoll, and J. E. Young, A closed-form differential renormalization-group generator for critical dynamics, *Phys. Lett.* 67A, p. 287, 1978.

Chang, T., S. W. Y. Tam, and C. C. Wu, Complexity induced anisotropic bimodal intermittent turbulence in space plasmas, *Phys. Plasmas* 11, p. 1287, 2004.

Chang, T., S. W. Y. Tam, and C. C. Wu, Complexity in space plasmas – a brief review, *Space Sci. Rev.* 122, p. 281, 2006.

Chang, T., D. D. Vvedensky, and J. F. Nicoll, Differential renormalization-group generators for static and dynamic critical phenomena, *Phys. Reports*, 217, p. 279, 1992.

Chang T., S. W. Y. Tam, C. C. Wu, and G. Consolini, Complexity, forced and/or self-organized criticality, and topological phase transitions in space plasmas. *Space Sci. Rev.* 107, p. 425, 2003.

Chang, T., C. C. Wu, J. Podesta, M. Echim, H. Lamy, and S. W. Y. Tam, ROMA (rank-ordered multifractal analyses) of intermittency in space plasmas – a brief tutorial review, *Nonlinear Processes in Geophysics* 17, p. 545, 2010.

Chapman, S. C., R. O. Dendy, and G. Rowlands, A sandpile model with dual scaling regimes for laboratory, space, and astrophysical plasmas. *Phys. Plasmas* 6, p. 4169, 1999.

Chapman, S. C., N. W. Watkins, R. O. Dendy, P. Helander, and G. Rowlands, A simple avalanche model as an analogue for the magnetospheric activity, *Geophys. Res. Lett.* 25, p. 2397, 1998.

Christensen, K., H. Flyvbjerg, and Z. Olami, Self-organized critical forest-fire model: Mean-field theory and simulation results in 1 to 6 dimensions, *Phys. Rev. Lett.* 71, p. 2737, 1993.

Clar, S., B. Drossel, and F. Schwabl, Scaling laws and simulation results for the self-organized critical forest-fire model, *Phys. Rev. E.* 50, p. 1009, 1994.

Consolini, G., Sandpile cellular automata and the magnetospheric dynamics, in *Cosmic Physics in the Year 2000*, edited by S. Aiello, N. Iucci, G. Sironi, A. Treves, and U. Villante, Proc VIII GIFCO Conference, Società Italiana di Fisica, Bologna, p. 123, 1997.

Consolini, G., and T. Chang, Magnetic field topology and criticality in geotail dynamics: Relevance to substorm phenomena, *Space Sci. Rev.* 95, p. 309, 2001.

Consolini, G., and T. Chang, Complexity, magnetic field topology, criticality, and metastability in magnetotail dynamics, *J. Atmos. Solar-Terrestrial Phys.* 64, p. 541, 2002.

Consolini, G., and P. De Michelis, A revised forest-fire cellular automaton for the nonlinear dynamics of the Earth's magnetosphere, *J. Atmos. Solar-Terrestrial Phys.* 63, p. 1371, 2001.

Consolini, G., and P. De Michelis, Rank ordering multifractal analysis of the auroral electrojet index, *Nonlinear Processes in Geophysics* 18, p. 277, 2011.

Consolini, G., and A. T. Y. Lui, Symmetry breaking and nonlinear wave-wave interaction in current disruption: Possible evidence for a phase transition, in *Magnetospheric Current Systems*, edited by S. Ohtani et al., Geophysical Monograph 118, AGU, Washington, DC, p. 395, 2000.

Consolini, G., T. Chang, and A. T. Y. Lui, Complexity and topological disorder in the Earth's magnetotail dynamics, in *Nonequilibrium Phenomena in Plasmas*, edited by A. S. Sharma, and P. K. Kaw, Springer, Dordrecht, The Netherlands, p. 51, 2005.

Consolini, G., M. F. Marcucci, and M. Candidi, Multifractal structure of auroral electrojet index data, *Phys. Rev. Lett.* 76, p. 4082, 1996.

Daróczy, Z., Generalized information functions, *Information and Control*, 16, p. 36, 1970.

Davis, T. N., and M. Sigiura, Aurora electrojet activity index AE and its universal time variations, *J. Geophys. R.* 71, p. 785, 1966.

Dennis, R. R., Solar hard X-ray bursts, *Solar Physics*, 100, p. 465, 1985.

Doe, R. A., J. D. Kelley, D. Lummerzheim, G. K. Parks, M. J. Brittnacher, G. A. Germany, and J. Spann, Initial comparison of POLAR UVI and Sondrestrom IS radar estimates for auroral electron energy flux, *Geophys. Res. Lett.* 24, p. 999, 1997.

Drossel, B., and F. Schwabl, Self-organized critical forest-fire model, *Phys. Rev. Lett.* 69, p. 1629, 1992.

Drossel, B., and F. Schwabl, Formation of space-time structure in a forest-fire model, *Physica A* 204, p. 212, 1994.

Drossel, B., S. Clar, and F. Schwabl, Exact results for the one-dimensional self-organized critical forest-fire model, *Phys. Rev. Lett.* 71, p. 3739, 1993.

Dubrulle, B., Intermittency in fully developed turbulence: Log-Poisson statistics and generalized scale invariance, *Phys. Rev. Lett.* 73, p. 959, 1994.

Echim, M. M., H. Lamy, and T. Chang, Multipoint observations of intermittency in the cusp regions, *Nonlinear Processes in Geophysics* 14, p. 525, 2007.

Einaudi, G., and M. Velli, The distribution of flares, statistics of magnetohydrodynamic turbulence and coronal heating, *Phys. Plasmas* 6, p. 4146, 1999.

Farge, M., Wavelet transforms and their applications to turbulence. *Ann. Rev. Fluid Mech.* 24, p. 395, 1992.

Feynman, J., and A. Ruzmaikin, Distributions of the interplanetary field revisited, *J. Geophys. Res.* 99, p. 17645, 1994.

Fisher, M. E., Correlation functions and the critical region of simple fluids, *J. Math. Phys.* 5, p. 944, 1964.

Fisher, M. E., Rigorous inequalities for critical-point correlation exponents, *Phys. Rev.* 180, p. 594, 1969.

Fisher, M. E., The theory of critical point singularities, in *Proc. 51st "Enrico Fermi" Summer School on Critical Phenomena* (Varenna, Italy), edited by M. S. Green, Academic, New York, 51, pp. 1–99, 1971.

Fisher, M. E., General scaling theory for critical points, in *Collective Properties of Physical Systems*, Proc. Nobel Symposium No. 24, edited by B. Lundqvist and S. Lundqvist, Academic Press, New York, p. 16, 1973.

Forman, M., and L. F. Burlaga, Exploring the Castaing distribution function to study intermittence in the solar wind at L1 in June 2000, in *Solar Wind Ten*, edited by M. Velli, R. Bruno, and F. Malara, *AIP Conf. Proc.* 679, p. 554, 2003.

Forster, D., D. R. Nelson, and M. J. Stephen, Large distance and long time properties of a randomly stirred fluid, *Phys. Rev. A* 16, p. 732, 1977.

Frisch, U., *Turbulence*, Cambridge University Press, Cambridge, 1995.

Georgoulis, M. K., M. Velli, and G. Einaudi, Statistical properties of magnetic activity in the solar corona, *Astrophys. J.* 497, p. 957, 1998.

Germany, G. A., M. R. Torr, and P. G. Richards, Use of FUV auroral emissions as diagnostics indicators, *J. Geophys. Res.* 99 (A1), p. 383, 1994.

Germany, G. A., G. K. Parks, M. Brittnacher, J. Cumnock, D. Lummerzheim, J. E. Spann, L. Chen, P. G. Richards, and F. J. Rich, Remote determination of auroral energy characteristics during substorm activity, *Geophys. Rev. Lett.* 24, p. 995, 1997.

Grassberger, P., Critical behavior of the Drossel-Schwabl forest-fire model, *New Journal of Physics* 4, p. 17.1, 2002.

Grassberger, P., and H. Kantz, On a forest-fire model with supposed self-organized criticality, *J. Stat. Phys.* 63, p. 685, 1991.

Griffiths, R. B., Thermodynamic inequality near the critical point for ferromagnets and fluids, *Phys. Rev. Lett.* 14, p. 623, 1965.

Grossmann, S., D. Lohse, and A. Reeh, Application of extended self-similarity in turbulence, *Phys. Rev. E* 56, p. 5473, 1997.

Haar, A., Zur theorie der orthogonalen functionensysteme, *Math. Ann.* 69, pp. 331–71, 1910.

Halperin, B. I., P. C. Hohenberg, and S. K. Ma, Calculation of dynamic critical properties using Wilson's expansion methods, *Phys. Rev. Lett.* 29, p. 1548, 1972.

Halsey, T., M. H. Jensen, L. P. Kadanoff, I. Procaccia, and B. I. Schraiman, Fractal measures and their singularities: The characterization of strange sets, *Phys. Rev. A* 33, p. 1141, 1986.

Heller, P., and Benedek, G. B., Nuclear magnetic resonance in MnF_2 near the critical point, *Phys. Rev. Lett.* 8, p. 428, 1962.

Hentschel, H. G. E., and I. Procaccia, The infinite number of generalized dimensions of fractals and strange attractors, *Physica D* 8, p. 435, 1983.

Hnat, B., S. C. Chapman, G. Rowlands, N. W. Watkins, and W. M. Farrell, Finite size scaling in the solar wind magnetic field energy density as seen by WIND, *Geophys. Res. Lett.* 29, p. 1446, doi:10.1029/2001GL014587, 2002.

Hnat, B., S. C. Chapman, G. Rowlands, N. W. Watkins, and M. P. Freeman, Scaling in long term data sets of geomagnetic indices and solar wind ε as seen by WIND spacecraft, *Geophys. Res. Lett.* 30, p. 2174, doi:10.1029/2003GL018209, 2003.

Horbury, T. A., A. Balogh, R. J. Forsyth, and E. J. Smith, Ulysses magnetic field observations of fluctuations within polar coronal flows. *Ann. Geophys.* 13, p. 105, 1995a.

Horbury, T. A., A. Balogh, R. J. Forsyth, and E. J. Smith, Observations of evolving turbulence in the polar solar wind. *Geophys. Res. Lett.* 22, p. 3401, 1995b.

Horbury, T. A., A. Balogh, R. J. Forsyth, and E. J. Smith, Ulysses observations of intermittent heliospheric turbulence, *Adv. Space. Phys.* 19, p. 847, 1997.

Hosino M., A. Nishida, T. Yamamoto, and S. Kokubun, Turbulent magnetic field in the distant magnetotail: Bottom-up process of plasmoid formation? *Geophys. Res. Lett.* 21, p. 2935, 1994.

Hwa, T., and M. Kardar, Avalanches, hydrodynamics, and discharge events in models of sandpiles. *Phys. Rev. A* 45, p. 7002, 1992.

Jensen, H. J., *Self-organized Criticality – Emergent Complex Behavior in Physical and Biological Systems*. Cambridge University Press, Cambridge, 1998.

Jensen H. J., K. Christenson, and H. C. Fogedby, 1/*f* noise, distribution of lifetimes, and a pile of sand, *Phys. Rev. B* 40, p. 7425, 1989.

Jiang, G. S., and C. C. Wu, A high order finite difference scheme for the equations of ideal magnetohydrodynamics, *J. Comp. Phys.* 150, p. 549, 1999.

Josephson, B. D., Inequality for the specific heat: II. Application to critical phenomena, *Proc. Phys. Soc.* 92, p. 276, 1967.

Kadanoff, L. P., Scaling laws for Ising models near T_c, *Physics* 2, p. 263, 1966.

Kadanoff, L. P., S. R. Nagel, L. Wu, and S. M. Zhou, Scaling and Universality in Avalanches, *Phys. Rev. A* 39, p. 6524, 1989.

Kadomtsev, B. B., and O. P. Pogutse, Nonlinear helical perturbations of a plasma in the tokamak, *Sov. Phys.-JETP* 38, p. 283, 1974.

Kaiser, A., and T. Schreiber, Information transfer in continuous processes, *Physica D* 166, p. 43, 2002.

Kertész, J., and L. B. Kiss, The noise spectrum in the model of self-organised criticality, *J. Phys. A (Math. and General)* 23, L433, 1990.

Kintner, P. M., J. Franz, P. Schunk, and E. Klatt, Interferometric coherency determination of wavelength or what are broadband ELF waves?, *J. Geophys. Res.* 105, p. 21237, 2000.

Klimas, A. J., V. M. Uritsky, and M. Paczuski, Self-organized criticality and intermittent turbulence in an MHD current sheet with a threshold Instability, *arXiv astro-ph:0701486v2*, 2007.

Klimas, A. J., V. M. Uritsky, D. Vassiliadis, and D. N. Baker, Reconnection and scale-free avalanching in a driven current-sheet model, *J. Geophys. Res.* 109, A02218, doi:10.1029/2003JA010036, 2004.

Klimas, A. J., V. M. Uritsky, D. Vassiliadis, and D. N. Baker, A mechanism for the loading-unloading substorm cycle missing in MHD global magnetospheric simulation models, *Geophys. Res. Lett.* 32, L14108, doi:10.1029/2005GL022916, 2005.

Klimas, A. J., D. N. Baker, D. A. Roberts, D. H. Fairfield, and J. Büchner, A nonlinear dynamical analogue model of geomagnetic activity. *J. Geophys. Res.* 97, p. 12253, 1992.

Klimas, A. J., J. A. Valdivia, D. Vassiliadis, D. N. Baker, M. Hesse, and J. Takalo, Self-organized criticality in the substorm phenomenon and its relation to localized reconnection in the magnetospheric plasma sheet, *J. Geophys. Res.* 105, p. 18765, 2000.

Lamy, H., A. Wawrzaszek, W. M. Macek, and T. Chang, New multifractal analyses of the solar wind turbulence: Rank-ordered multifractal analysis and generalized two-scale weighted Cantor set model, in *Solar Wind 12*, edited by M. Maksimovic, K. Issautier, N. Meyer-Vernet, M. Moncuquet, and F. Pantellini, Twelfth International Solar Wind Conference, AIP Conference Series, 1216, p. 124, 2010.

Lamy, H., M. Echim, and T. Chang, Rank-ordered multifractal spectrum of intermittent fluctuations in the cusp: A case study with Cluster data, 37th COSPAR Scientific Assembly, Paper D31-0017-08, p. 1686, 2008.

Landau, L. D., and E. M. Lifshitz, *Statistical Physics*, Addison-Wesley, Reading, MA, 1970.

Leubner, M. P., Fundamental issues on kappa-distributions in space plasmas and interplanetary proton distributions, *Phys. Plasmas* 11, p. 1308, 2004.

Leubner, M. P., A nonextensive entropy approach to kappa-distributions, *Astrophysics and Space Science* 282, p. 573, 2002.

Leubner, M. P., Fundamental issues on kappa-distributions in space plasmas and interplanetary proton distributions, *Phys. Plasmas* 11, p. 1308, 2004.

Leubner, M. P., and Z. Vörös, A nonextensive entropy approach to solar wind intermittency, *Astrophys. J.* 618, p. 547, 2005.

Li, Y., E. Perlman, M. Wan, Y. Yang, C. Meneveau, R. Burns, S. Chan, A. Szalay, and G. Eyink, A public turbulence database cluster and applications to study Lagrangian evolution of velocity increments in turbulence, *J. Turbulence* 9, p. 1, 2008.

Loreto, V., A. Vespignani, and S. Zapperi, Renormalization scheme for forest-fire models, *J. Phys. A, Math. Gen.* 29, p. 2981, 1996.

Loreto, V., L. Pietronero, A. Vespignani, and S. Zapperi, Renormalization-group approach to the critical behavior of the forest-fire model, *Phys. Rev. Lett.* 75, p. 465, 1995.

Lu, E. T., Avalanches in continuum driven dissipative systems. *Phys. Rev. Lett.*, 74, p. 2511, 1995.

Lu, E. T., and R. J. Hamilton, Avalanches and the distribution of solar flares, *Astrophys. J.* 380, L89–L92, 1991.

Lui, A. T. Y., Current disruption in the Earth's magnetosphere: observations and models, *J. Geophys. Res.* 101, p. 13067, 1996.

Lui, A. T. Y., Multifractal and intermittent nature of substorm-associated magnetic turbulence in the magnetotail, *J. Atmos. Solar-Terrestrial Phys.* 63, p. 1379, 2001.

Lui, A. T. Y., Multiscale phenomena in the near-Earth magnetosphere, *J. Atmos. Solar-Terrestrial Phys.* 64, p. 125, 2002.

Lui, A. T. Y., C. L. Chang, A. Mankofsky, H. K. Wong, and D. W inske, A cross-field current instability for substorm expansions, *J. Geophys. Res.* 96, p. 11389, 1991.

Lui, A. T. Y., R. E. Lopez, S. M. Krimigis, R. W. McEntire, L. J. Zanetti, and T. A. Potemra, A case study of magnetotail current sheet disruption and diversion, *Geophys. Res. Lett.* 15, p. 721, 1988.

Lui, A. T. Y., S. C. Chapman, K. Liou, P. T. Newell, C. I. Meng, M. Brittnacher, and G. K. Parks, Is the dynamic magnetosphere an avalanching system? *Geophys. Res. Lett.* 27, p. 911, 2000.

Lui, A. T. Y., K. Liou, P. T. Newell, C. I. Meng, S. I. Ohtani, S. Kokubun, T. Ogino, M. Brittnacher, and G. Parks, Plasma and magnetic flux transport associated with auroral breakups, *Geophys. Res. Lett.* 25, p. 4059, 1998.

Lui, A. T. Y., Y. Zheng, Y. Zhang, H. Rème, M. W. Dunlop, G. Gustafsson, S. B. Mende, C. Mouikis, and L. M. Kistler, Cluster observation of plasma flow reversal in the magnetotail during a substorm, *Ann. Geophys.* 24, p. 2005, 2006.

Macek, W. M., Multifractality and intermittency in the solar wind, *Nonlinear Processes in Geophysics*, 14, p. 695, 2007.

Macek, W. M., and A. Wawrzaszek, Evolution of asymmetric multifractal scaling of solar wind turbulence in the outer heliosphere, *J. Geophys. Res.*, 114, A031108, doi:10.1029/2008JA013795, 2009.

Macek, W. M., and A. Wawrzaszek, Multifractal structure of small and large scales fluctuations of interplanetary magnetic fields, *Planet. Space Sci.* 59, p. 569, 2011a.

Macek, W. M., and A. Wawrzaszek, Multifractal two-scale Cantor set model for slow solar wind turbulence in the outer heliosphere during solar maximum, *Nonlinear Processes in Geophysics* 18, p. 287, 2011b.

Macek, W. M., A. Wawrzaszek, and L. F. Burlaga, Multifractal structures detected by Voyager 1 at the heliospheric boundaries, *Ap. J. Letters*, 793, L30, 2014.

Macek, W. M., A. Wawrzaszek, and V. Carbone, Observation of the multifractal spectrum at the termination shock by Voyager 1, *Geophys. Res. Lett.* 38, L19103 2011.

Macek, W. M., A. Wawrzaszek, and V. Carbone, Observation of the multifractal spectrum in the heliosphere and the heliosheath by Voyager 1 and 2, *J. Geophys. Res.* 117, A12101, doi:10.1029/2012JA018129, 2012.

Mallet, S., and W. L. Huang, Singularity detection and processing with wavelets, *IEEE Trans. on Information Theory* 38, p. 617, 1992.

Mallet, S., and S. Zhong, Characterization of signals from multiscale edges, *IEEE Trans. on Pattern Analysis and Machine Intelligence* 14, p. 710, 1992.

Mandelbrot, B., *Fractals: Form, Chance and Dimension*, Freeman & Co., San Francisco, 1977.

Marsch, E., and S. Liu, Structure functions and intermittency of velocity fluctuations in the inner solar wind, *Ann. Geophys.* 11, p. 227, 1993.

Marsch, E., and C. Y. Tu, Non-Gaussian probability distributions of solar wind fluctuations, *Ann. Geophys.* 12, p. 1127, 1994.

Marsch, E., and C. Y. Tu, Intermittency, non-Gaussian statistics and fractal scaling of MHD fluctuations in the solar wind, *Nonlinear Processes in Geophysics* 4, p. 101, 1997.

Marsch E., C. Y. Tu, and H. Rosenbauer, Multifractal scaling of the kinetic energy flux in solar wind turbulence, *Ann. Geophys.* 14, p. 259, 1996.

Matthaeus, W. H., and M. L. Goldstein, Stationarity of magnetohydrodynamic fluctuations in the solar wind, *J. Geophys. Res.* 87, p. 10347, 1982.

Matthaeus, W. H., M. L. Goldstein, and D. A. Roberts, Evidence for the presence of quasi-two-dimensional nearly incompressible fluctuations in the solar wind, *J. Geophys. Res.* 95, p. 20673, 1990.

Matthaeus, W. H., G. Qin, J. W. Bieber, and G. P. Zank, Nonlinear collisionless perpendicular diffusion of charged particles, *Astrophys. J. Lett.* 590, L53, 2003.

Meneveau, C., Analysis of turbulence in the orthonormal wavelet representation. *J. Fluid Mech.* 232, p. 469, 1991.

Meneveau, C., and K. R. Sreenivasan, Simple multifractal cascade model for fully developed turbulence, *Phys. Rev. Lett.* 59, p. 1424, 1987.

Milovanov, A. V, L. M. Zelenyi, and G. Zimbardo, Fractal structures and power law spectra in the distant Earth's magnetotail, *J. Geophys. Res.* 101, p. 19903, 1996.

Morlet, J., Sampling theory and wave propagation, in *Issues on Acoustic Signal/Image Processing and Recognition*, edited by C. H. Chen, NATO ASI, F1, p. 233, Springer, Berlin, 1983.

Mossner, W. K., B. Drossel, and F. Schwabl, Computer simulations of the forest-fire model, *Physica A* 190, p. 205, 1992.

Müller, W. C., and D. Biskamp, Scaling properties of three-dimensional magnetohydrodynamic turbulence, *Phys. Rev. Lett.* 84, p. 475, 2000.

Muzy, J. F., E. Bacry, and A. Arnéodo, The multifractal formalism revisited with wavelets, *Intern. J. of Bifurcation and Chaos* 4, p. 245, 1994.

Ott, E., *Chaos in Dynamical Systems*, Cambridge University Press, New York, 1993.

Pagel, C., and A. Balogh, A study of magnetic fluctuations and their anomalous scaling in the solar wind: the Ulysses fast-latitude scan, *Nonlinear Processes in Geophysics* 8, p. 313, 2001.

Parisi, G., and U. Frisch, On the singularity structure of fully developed turbulence, in *Turbulence and Predictability in Geophysical Fluid Dynamics*, edited by M. Ghil, R. Benzi, and G. Parisi, Proc Intern School of Physics, "Erico Fermi" 1983, Varenna, Italy, North Holland, Amsterdam, p. 84, 1985.

Perlman, E., R. Burns, Y. Li, and C. Meneveau, Data exploration of turbulence simulations using a database cluster, Article No. 23, Supercomputing SC07, *ACM, IEEE*, doi:10.1145/1362622, 1362654, 2007.

Petviashvili, V. I., and O. A. Pokhotelov, *Solitary Waves in Plasmas and in the Atmosphere*, Gordon and Breach Science Publishers, Reading, MA, 1992.

Pietronero, L., A. Vespignani, and S. Zapperi, Renormalization scheme for self-organized criticality in sandpile models, *Phys. Rev. Lett.* 72, p. 1690, 1994.

Politano, H., and A. Pouquet, Model of intermittency in magnetohydrodynamic turbulence, *Phys. Rev. E*, 52, p. 636, 1995.

Politano, H., A. Pouquet, and V. Carbone, Determination of anomalous exponents of structure functions in two-dimensional magnetohydrodynamic turbulence, *Europhys. Lett.* 43, p. 516, 1998.

Press, W. H., Flicker noise in astronomy and elsewhere, *Comments on Modern Physics* (Part C: Comments on Astrophysics) 7, p. 103, 1978.

Pruessner, G., and H. J. Jensen, Broken scaling in the forest-fire model, *Phys. Rev. E* 65, 056707, 2002.

Pruessner, G., and H. J. Jensen, Efficient algorithm for the forest-fire model, *Phys. Rev. E* 70, 066707, 2004.

Roberts, D. A., D. N. Baker, A. J. Klimas, and L. F. Bargatze, Indications of low dimensionality in magnetospheric dynamics. *Geophys. Res. Lett.* 18, p. 151, 1991.

Rushbrooke, G. S., On the thermodynamics of the critical region for the Ising problem, *J. Chem. Phys.* 39, p. 842, 1963.

Rusmaikin, A., I. R. Lyannaya, V. A. Styashkin, and E. Yeroshenko, The spectrum of the interplanetary magnetic field near 1.3 AU, *J. Geophys. Res.* 98, p. 13303, 1993.

Seyler, C. E., A mathematical model of the structure and evolution of small-scale discrete auroral arcs, *J. Geophys. Res.* 95, p. 17199, 1990.

Shan L. H., C. K. Goertz, and R. A. Smith, On the embedding-dimension analysis of AE and AL time series. *Geophys. Res. Lett.* 18, p. 1647, 1991b.

Shan, L. H., P. Hansen, C. K. Goertz, and R. A. Smith, Chaotic appearance of the AE index. *Geophys. Res. Lett.* 18, p. 147, 1991a.

Shannon, C. E., A mathematical theory of communication, *The Bell System Technical Journal* 27, p. 379 and p. 623, 1948.

Sharma, A. S, D. V. Vassiliadis, and K. Papadopoulos, Reconstruction of low-dimensional magnetospheric dynamics by singular spectrum analysis. *Geophys. Res. Lett.* 20, p. 335, 1993.

She, Z.-S., and E. Lévêque, Universal scaling laws in fully developed turbulence, *Phys. Rev. Lett.* 72, p. 336, 1994.

She, Z.-S., and E. C. Waymire, Quantized energy cascade and log-Poisson statistics in fully developed turbulence, *Phys. Rev. Lett.* 74, p. 262, 1995.

Sornette, D., *Critical Phenomena in Natural Sciences, Chaos, Fractals, Self-organization and Disorder: Concepts and Tools*, Springer, Heidelberg, 2000.

Sorriso-Valvo, L., V. Carbone, P. Veltri, G. Consolini, and R. Bruno, Intermittency in the solar wind turbulence through probability distribution functions of fluctuations, *Geophys. Res. Lett.* 26, p. 1801, 1999.

Sorriso-Valvo, L., V. Carbone, P. Giuliani, P. Veltri, R. Bruno, V. Antoni, and E. Martines, Intermittency in plasma turbulence, *Planetary Space. Sci.* 49, p. 1193, 2001.

Stanley, H. E., *Introduction to Phase Transitions and Critical Phenomena*, Oxford University Press, New York, 1971.

Stasiewicz, K., Y. Khotyaintsev, M. Berthomier, and J.-E. Wahlund, Identification of widespread turbulence of dispersive Alfvén waves, *Geophys. Res. Lett.* 27, p. 173, 2000.

Strauss, H. R., Nonlinear, three-dimensional magnetohydrodynamics of noncircular tokamaks, *Phys. Fluids*, 19, p. 134, 1976.

Sundkvist, D., A. Retinò, A. Vaivads, and S. D. Bale, Dissipation in turbulent plasma due to reconnection in thin current sheets, *Phys. Rev. Lett.* 99, 025004, 2007.

Sundkvist, D., V. Krasnoselskikh, P. K. Shukla, A. Vaivads, M. André, S. Buchert, and H. Rème, In situ multi-satellite detection of coherent vortices as a manifestation of Alfvénic turbulence,

Nature 436, p. 825, 2005; erratum, 437, p. 290, 2005.

Takalo, J., Correlation dimension of AE data, Ph. Lic. Thesis, Laboratory Report 3, Department of Physics, University of Jyväskylä, 1993.

Takalo, J., J. Timonen, A. Klimas, J. Valdivia, and D. Vassiliadis, A coupled-map model for the magnetotail current sheet, *Geophys. Res. Lett.* 26, p. 2913, 1999.

Takalo, J., J. Timonen, A. Klimas, J. Valdivia, and D. Vassiliadis, A coupled-map as a model of the dynamics of the magnetotail current sheet, *J. Atmos. Solar-Terrestrial Phys.* 63, p. 1407, 2001.

Tam, S. W. Y., T. Chang, G. Consolini, and P. De Michelis, Renormalization-group description and comparison with simulation results for forest-fire models – possible interpretation of near-criticality phenomenon in the dynamics of space plasmas, *Eos Trans. AGU* 81 (48), Fall Meet. Suppl., Abstract SM62A-04, 2000.

Tam, S. W. Y., and T. Chang, Energization of ions by bimodal intermittent fluctuations, in *Multiscale Coupling of Sun-Earth Processes*, edited by A. T. Y. Lui, Y. Kamide, and G. Consolini, Elsevier, The Netherlands, p. 375, 2005.

Tam, S. W. Y., and T. Chang, Double rank-ordering technique of ROMA (rank-ordered multifractal analysis) for multifractal fluctuations featuring multiple regimes of scales, *Nonlinear Processes in Geophysics* 18, p. 405, 2011.

Tam, S. W. Y., T. Chang, P. M. Kintner, and E. Klatt, Intermittency analyses on the SIERRA measurements of the electric field fluctuations in the auroral zone, *Geophys. Res. Lett.* 32, L05109, doi:10.1029/2004GL021445, 2005.

Tam, S. W. Y., T. Chang, P. M. Kintner, and E. Klatt, Rank-ordered multifractal analysis for intermittent fluctuations with global crossover behavior, *Phys. Rev. E*, 81, 036414, 2010.

Tanaka, M., Meaning of an escort distribution and τ- transformation, in *Mathematical Aspects of Generalized Entropies and Their Applications, Journal of Physics: Conference Series* 201, 012007, 2010.

Taylor, G. I., The spectrum of turbulence, *Proc. Royal Soc. London A* 164, p. 476, 1938.

Taylor, J. B., Relaxation of toroidal plasma and generation of reverse magnetic fields, *Phys. Rev. Lett.* 33, p. 1139, 1974.

Temerin, M., Doppler shift effects on double-probe-measured electric field power spectra, *J. Geophys. Res.* 84, p. 5929, 1979.

Tetreault, D., Turbulent relaxation of magnetic fields. 1. Coarse-grained dissipation and reconnection, *J. Geophys. Res.* 97, p. 8531, 1992a.

Tetreault, D., Turbulent relaxation of magnetic fields. 2. Self-organization and intermittency, *J. Geophys. Res.* 97, p. 8541, 1992b.

Torrence, C., and G. P. Compo, A practical guide to wavelet analysis (available online: http://paos.colorado.edu/research/wavelets), *Bulletin of the American Meteorological Society* 79, p. 61, 1998.

Tsallis, C., Possible generalization of Boltzmann-Gibbs statistics, *J. Stat. Phys.* 52, p. 479, 1988.

Tsallis, C., R. S. Mendes, and A. R. Plastino, The role of constraints within generalized nonextensive statistics, *Physica A* 261, p. 534, 1998.

Tsurutani, B. T., M. Sugiura, T. Iyemori, B. E. Goldstein, W. D. Gonzalez, S. I. Akasofu, and E. J. Smith, The nonlinear response of AE to the IMF B_s Driver: a spectral break at 5 hours, *Geophys. Res. Lett.* 17, p. 279, 1990.

Tu, C. Y., and E. Marsch, *MHD Structures, Waves and Turbulence in the Solar Wind – Observations and Theories*, Kluwer Academic Publishers, Dordrecht, The Netherlands, 1995.

Tu, C. Y., E. Marsch, and H. Rosenbauer, An extended structure function model and its application to the analysis of solar wind intermittency properties, *Ann. Geophys.* 14, p. 270, 1996.

Turcotte, D. L., *Fractals and Chaos in Geology and Geophysics*, Cambridge University Press, New York, 1997.

Uritsky, V. M., and A. J. Klimas, Hysteresis-controlled instability waves in a scale-free driven current sheet model. *Nonlinear Processes in Geophysics* 12, p. 827, 2005.

Uritsky, V. M., A. J. Klimas, and D. Vassiliadis, Multiscale dynamics and robust critical scaling in a continuum current sheet model, *Phys. Rev. E* 65, 046113, 2002b.

Uritsky, V. M., A. J. Klimas, and D. Vassiliadis, Critical finite-size scaling of energy and lifetime probability distributions of auroral emissions, *Geophys. Res. Lett.* 33, L08102, doi:10.1029/2005GL025330, 2006.

Uritsky, V. M., M. Pudovkin, and A. Steen, Geomagnetic substorms as perturbed self-organized critical dynamics of the magnetosphere, *J. Atmos. Solar-Terrestrial Phys.* 63, p. 1415, 2001a.

Uritsky, V. M., M. Paczuski, J. M. Davila, and S. I. Jones, Coexistence of self-organized criticality and intermittent turbulence in the solar corona. *Phys. Rev. Lett.* 99, 025001, 2007.

Uritsky, V. M., A. J. Klimas, J. A. Valdivia, D. Vassiliadis, and D. N. Baker, Stable critical behavior and fast field annihilation in a magnetic field reversal model, *J. Atmos. Solar-Terrestrial Phys.* 63, p. 1425, 2001b.

Uritsky, V. M., A. J. Klimas, D. Vassiliadis, D. Chua, and G. Parks, Scale-free statistics of spatiotemporal auroral emissions as depicted by POLAR UVI images: Dynamic magnetosphere is an avalanching system. *J. Geophys. Res.* 107, A12, p. 1426, doi: 10.1029/2001JA000281, 2002a.

Vahnstein, S., K. Sreenivasan, R. Pierrehumbert, V. Kashyap, and A. Juneja, Scaling exponents for turbulence and other random processes and their relationships with multifractal structure, *Phys. Rev. E* 50, p. 1823, 1994.

Vassiliadis, D., A. Anastasiadis, M. Georgoulis, and L. Vlahos, Derivation of solar flare cellular automata models from a subset of the magnetohydrodynamic equations, *Astrophys. J.* 509, L53, 1998.

Vasyliunas, V. M., A Survey of low-energy electrons in the evening sector of the magnetosphere with OGO 1 and OGO 3, *J. Geophys. Res.*, 73, p. 2839, 1968.

Vespignani, A., and S. Zapperi, Order parameter and scaling fields in self-organized criticality, *Phys. Rev. Lett.* 78, p. 4793, 1997.

Vespignani, A., and S. Zapperi, How self-organized criticality works: A unified mean-field theory, *Phys. Rev. E* 57, p. 6345, 1998.

Vespignani, A., S. Zapperi, and L. Pietronero, Renormalization approach to the self-organized critical behavior of sandpile models, *Phys. Rev. E* 51, p. 1711, 1995.

Vlahoes, L., M. Georgoulis, R. Kluiving, and P. Paschos, The statistical flare, *Astronomy and Astrophysics.* 299, p. 897, 1995.

Vörös, Z., On multifractality of high-latitude geomagnetic fluctuations, *Ann. Geophys.* 18, p. 1273, 2000.

Vörös Z., W. Baumjohann, R. Nakamura, A. Runov, M. Volwerk, T. L. Zhang, and

A. Balogh, Wavelet analysis of magnetic turbulence in the Earth's plasma sheet, *Phys. Plasmas* 11, p. 1333, 2004.

Vörös, Z., W. Baumjohann, R. Nakamura, A. Runov, M. Volwerk, H. Schwarzl, A. Balogh, and H. Rème, Dissipation scales in the Earth's plasma sheet estimated from Cluster measurements, *Nonlinear Processes in Geophysics* 12, p. 725, 2005.

Vörös, Z., W. Baumjohann, R. Nakamura, A. Runov, T. L. Zhang, M. Volwerk, H. U. Eichelberger, A. Balogh, T. S. Horbury, K. H. Glassmeier, B. Klecker, and H. Rème, Multi-scale magnetic field intermittence in the plasma sheet, *Ann. Geophys.* 21, p. 1955, 2003.

Watkins, N. W., S. C. Chapman, R. O. Dendy, and G. Rowlands, Robustness of collective behavior in strongly driven avalanche models: magnetospheric implications. *Geophys. Res. Lett.* 26, p. 2617, 1999.

Weygand, J. M., M. G. Kivelson, K. K. Khurana, H. K. Schwarzl, S. M. Thompson, R. L. McPherron, A. Balogh, L. M. Kistler, M. L. Goldstein, J. Borovsky, and D. A. Roberts, Plasma sheet turbulence observed by Cluster II, *J. Geophys. Res.* 110, A01205, doi:10.1029/2004JA010581, 2005.

Wilson, K. G., Renormalization group and critical phenomena. I. Renormalization group and the Kadanoff scaling picture, *Phys. Rev. B* 4, p. 3174, 1971a.

Wilson, K. G., Renormalization group and critical phenomena. II. Phase-space cell analysis of critical behavior, *Phys. Rev. B* 4, p. 3184, 1971b.

Wilson, K. G., and J. Kogut, The renormalization group and the ε expansion, *Physics Reports* 12, p. 75, 1974.

Wu, C. C., and T. Chang, 2D MHD simulation of the emergence and merging of coherent structures, *Geophys. Res. Lett.* 27, p. 863, 2000.

Wu, C. C., and T. Chang, Further study of the dynamics of two-dimensional MHD coherent structures – a large scale simulation, *J. Atmos. Solar-Terrestrial Phys.* 63, p. 1447, 2001.

Wu, C. C., and T. Chang, Intermittent turbulence in 2D MHD simulation, in *Multiscale Coupling of Sun-Earth Processes*, edited by A. T. Y. Lui, Y. Kamide, and G. Consolini, Elsevier, The Netherlands, p. 321, 2005.

Wu, C. C., and T. Chang, Rank-ordered multifractal analysis (ROMA) of probability distributions in fluid turbulence, *Nonlinear Processes in Geophysics* 18, p. 261, 2011.

Wu, C. C., T. Chang, and V. M Uritsky, Rank-ordered multifractal analysis (ROMA) of intermittent dissipative structures in solar corona, Abstract SH51C-2020, 2011 Fall Meeting, AGU, San Francisco.

Yordanova, E., M. Grezesiak, A. W. Wernik, B. Popielawska, and K. Stasiewicz, Multifractal structure of turbulence in the magnetospheric cusp, *Ann. Geophys.* 22, p. 2431, 2004.

Yordanova, E., J. Bergman, G. Consolini, M. Kretzschmar, M. Materassi, B. Popielawska, M. Roca-Sogorb, K. Stasiewicz, and A. W. Wernik, Anisotropic scaling features and complexity in magnetospheric-cusp: a case study, *Nonlinear Processes in Geophysics* 12, p. 817, 2005.

Index